Darts on History of Mathematics

SATISH C. BHATNAGAR

BOOKS BY THE SAME AUTHOR

1. Scattered Matherticles: Mathematical Reflections, Volume I
2. Vectors in History: Main Foci - India and USA, Volume 1
3. Epsilons of Deltas of Life: Everyday Stories, Volume I
4. My Hindu Faith & Periscope, Volume I
5. Via Bhatinda: A Braid of Reflected Memoirs, Volume I
6. Swami Deekshanand Saraswati: My Swami Mama Ji

Order this book online at www.trafford.com
or email orders@trafford.com

Most Trafford titles are also available at major online book retailers.

Printed in the United States of America.

ISBN: 978-1-4907-4937-2 (sc)
ISBN: 978-1-4907-4938-9 (e)

Trafford rev. 10/28/2014

 www.trafford.com

North America & international
toll-free: 1 888 232 4444 (USA & Canada)
fax: 812 355 4082

SELECTED COMMENTS

This is a wonderful testament to a balanced academic life consisting of teaching and research, and sharing our passion for education and training the next generation of leadership....it is certainly worth a departmental discussion. **Neal Smatresk** (former UNLV President, currently of North Texas - on # 15)

I just wanted to reply to this particular reflection since I really take it to heart. I do try to read all the reflections that you e-mail to us students, but I have to admit that I actually made it a point to read this one several times. I especially love the following sentences: "A professional paradox is that good students are never inspired by instructors whose hearts are not in teaching. And, without a nursery of excellent students, a good crop of researchers can never be expected."

As I read this reflection, I had to stop at the end of that first sentence and read it out loud to fully enjoy it. I have wanted to be a teacher for as long as I can remember and I think the highest compliment I ever receive from students is when they tell me they can see that I really enjoy teaching. Sometimes I feel a little overwhelmed by my teaching duties, but from now on I'm going to remember what you've written in this reflection. I will always try to remember that my heart is in teaching and that in order to inspire my students, I need to try not to hide it. So thank you. - **Megan Austin** (Currently CSN Instructor - on # 17)

Satish, I'm glad that you've seen the light and become a historian!! Seriously, this sounds very intriguing. **Michael Bowers** (UNLV Provost and Executive VP –on #21)

I think my B. Banneker award recipients would appreciate a copy of that reflection, do you mind sending it to me as an attachment? They were all impressed I had a course from you! Thanks, **Aaron** on # 29 (About to get PhD in math education)

Your reflections are inspiring and often infuse new spirit in me. I carried them with me when I left for holiday and were part of my breakfast. **Ahmed Yagi** (Associate Dean UN, Oman - on # 34)

Your account of a century of mathematics in Punjab is extremely well written and perceptive. **Arun Vaidya**, (Retired Prof and Head of Math Dept, Gujarat University, Ahmedabad – on # 43)

Satish, that was one of the best of your always interesting essays. A lot of ideas for further exploration. This is not an area I give much thought to because I have little respect for religion and its impact on the world, but that is a different issue. Take care. Ciao, **Len Zane** Emeritus Professor of Physics on # 55

Darts on History of Mathematics

DEDICATED TO

"HISTORY FOR THE WISE"

SATISH C. BHATNAGAR

DISTRIBUTION OF CONTENTS

EXTRACTS FROM THE PREFACE OF *VECTORS IN HISTORY*......... 1
A HIST-O-MATH PREFACE.. 5
DESIGNING THE FRONT COVER...11
GLOSSARY AND ABBREVIATONS 13

I. CLASSROOMS CUTS... 15

    2.   CALCULUS DEFINES CIVILIZATIONS 17
    3.   ON HISTORY OF MATHEMATICS COURSE.................... 19
    4.   MY PERCEPTION OF HISTORY IN GENERAL 22
    5.   HANDS-ON HISTORY 26
    6.   HISTORY OF/IN MATHEMATICS.......................................28
    7.   CONDUCTING HISTORY PROJECTS 31
    8.   A SAMPLE LETTER FOR HELPING STUDENTS 33
    9.   PUSH - PROJECT - PROGRESS 34
    10. RATIONALE FOR HISTORY OF MATH COURSE 36
    11. A MATHEMATICAL EXCITEMENT 38
    12. A PUDDING PROOF!... 43
    13. CHALLENGES OF FINDING FACTS 45
    14. A SAGA OF ELEMENTARY CALCULUS......................... 49
    15. DISCOVERING MATHEMATICAL HISTORY 52
    16. EXTRA CREDITS ON MATHEMATICAL HISTORY........ 56
    17. NIZWA WITHOUT HISTORY! 58
    18. DARK SPOTS ON THE MOONS 60
    19. TEACHING BY THE RE-SEARCHERS...............................62
    20. FIELD TRIP & HISTORY OF MATHEMATICS..................64
    21. THE FALL AND RISE OF TEACHING..............................66
    22. 'SOME' HISTORY OF MATHEMATICS 71
    23. WRITINGS ON IIISTORY OF MATHEMATICS 75
    24. AN APPETIZER IN HISTORY 78
    25. DESSERTS IN A HISTORY COURSE 81
    26. UNEXPECTED HISTORICAL NUGGETS 84

II. HUMANISTIC SLICES 87

    27. ON INTERVIEW WITH MARY RUDIN 89
    28. LAS VEGAS & SAUNDERS MACLANE 91
    29. A TALE OF TWO MATHEMATICS 94
    30. WHO IS A GREAT MATHEMATICIAN? 99
    31. MARBLES OF RESEARCH 102
    32. P. R. HALMOS, AS I REMEMBER 105
    33. BENJAMIN BANNEKER AND MATHEMATICS 112
    34. GONE! GO GOLBERG 115
    35. ON MATHEMATICAL IMMORTALITY 118
    36. INVITING FACULTY INTO RESEARCH 124
    37. WE NEARLY MISSED EACH OTHER! 126
    38. RETURN OF A KIND! 131
    39. GIANTS IN THE CLASSROOM! 134
    40. DARK SPOTS ON THE MOONS (PART II) 138
    41. CONNECTING THE DOTS 143

III. INDIAN SPICES 147

    42. GENERAL MATHEMATICS IN THE VEDAS 149
    43. MATHEMATICS AND COLONIZATION 154
    44. HISTORY, MATHEMATICS & HERITAGE 159
    45. NEW MATERIALS ON HISTORY OF MATHEMATICS ..161
    46. NUMBER-LESS-NESS OF NUMBERS! 164
    47. HISTORY OF MATHEMATICS IN PUNJAB 168
    48. CULTURE - CONFERENCE COMBINE 178
    49. DECKING OUT HINDU MATHEMATICIANS 183

IV. SMORGASBORD BITES 191

    50. PHILOSOPHY TURNING MATHEMATICAL 193
    51. MATHEMATICS AND HINDU *DHARMA* 198
    52. HOLOCAUST AND GODEL'S THEOREM 204
    53. A CONVERGENCE OF SLAVERY AND
        MATHEMATICS 208
    54. GAUSSIAN CHAIR 210
    55. PERU (INCA CIVILIZATION): A PERISCOPE 213

56. WOMEN, MATHEMATICS AND HINDUISM 221
57. MATHEMATICS, SCIENCE AND RELIGIONS 225
58. STIMULUS OF THE FIRST LOVE 228
59. BRUSHING OVER THEOLOGY AND MATHEMATICS .. 231
60. HISTORY, MATHEMATICS & HERITAGE 239
61. ON AN HISTORIC IF 'N BUT 246
62. A CHALLENGE MET! 248
63. INVERSE PROBLEMS IN ARCHAELOGICAL
      MATHEMATICS 252
64. ICM: A MATHEMATICANS' SHOW 257
65. MATHEMATICIANS IN MANHATTAN PROJECT?? 260

V. OTHER PERSPECTIVES 265

66. WHAT HOM MEANS TO YOU! 267
67. ORTHOGONALITY OF MATHEMATICS & HISTORY .. 269
68. MANGHO AHUJA 272
69. ALOK KUMAR 275
70. ADAM KOEBKE 279
71. SHAODONG LIN 280
72. OWEN NELSON 282

COMMENTATORS & ANALYSTS EXTRAORDINAIRE 285

EXTRACTS FROM THE PREFACE
OF *VECTORS IN HISTORY*

A rationale for including the following extracts from **VECTORS IN HISTORY** is that the **DARTS ON HISTORY OF MATHEMATICS** is its 'mathematical' corollary. My approach to history of mathematics is holistic in a sense that I consider history of mathematics as a subset of history in general. Therefore, establishing my credentials as historian was a must. Having read scores of history books is not enough. Nevertheless, I do pass the 10,000 hour test!

However, it is a sense of history in interpreting past and present events and projecting them in future that underlines the making of a historian. In humanities and social studies, writing a book earns tenure and promotion– though, one may still be far from being an historian.

"In India, I have known people getting master's in history by 'mugging up' a dozen questions from various 'guess papers', openly sold in the market. By Indian standards, it was still considered laudable, but by the US academic standards, such master's degrees are jokes. There is no development of any sense of history of events, leaders, movements and ideas. They could never connect the dots and interpret events. But they earn all professional perks and benefits of master's degrees.

"In the US, over the years, the works of the likes of Will and Ariel Durant have deeply influenced me, as I enjoyed writing my articles and letters for daily newspapers and weeklies. The real breakthrough came when I started listening to radio talk shows – while driving back and forth to work. I was stunned by hosts' knowledge of history, connections with the current issues, power of communication, and a following of 5-25 million people, per week, across the USA.

"The word 'Vector', in the title of the book, VECTORS IN HISTORY, is borrowed from mathematics, where vectors are not restricted to only two or three dimensions, as used in physics and geometry. Vectors represent quantities that need both magnitude and direction for their

representations. Popularly, vectors are shown by directed arrows 'freely flying around'.

"Each Reflection in this book is, indeed, like a vector. Its magnitude is subjective, if measured by its impact factor - but tangible, if it is measured by the number of words - varying from 600 to 6000. Its theme provides a direction; however, it may change its course during its ricocheting flight. In other words, a particular Reflection may have more than one attractor points.

"In mathematics and physics, two position vectors/forces can be added. In Reflections, there is generally no connection between the ending of one Reflection and beginning of the next. Consequently, one can read a Reflection from anywhere without missing a beat from the previous one! That is a kind of beauty and uniqueness of this book.

"A general structure of a Reflection is that it spins off from a specific incident - whether in India or the USA. I basically dive into a vortex with it, or look out through it as a window to the universe far and beyond. In the process, it is wrapped with other stories and concepts which are often cross-cultural and inter-disciplinary. The delight they have given me on subsequent readings would surely give in different measures to the readers.

"Each of the 79 Reflections contains at least one nugget, though it may not glitter equally for everyone. For the last six months, while intensely sorting, compiling, and editing them, these Reflections gave me moments of new thrill and gratification.

"My mind is becoming a single track that I can't work on two Reflections side by side. Once, a new Reflection is started off, the previous one is literally expelled out of the mind. The book has something for everyone – including students, laypersons and professionals. It may be used as a supplementary book for any course on history of India from a global perspective. Above all, its purpose is to ignite interest in general history too.

"These days a long bibliography is interpreted as a sign of scholarship in arts, humanities and social studies. Recently, I confronted a

person about this when some works were never even looked at. In my experience, over the years, bibliographies - like intellectual mafia, may be perpetuating myths or stretching facts of a particular school of thought.

"For instance, the British colonial historians systematically created numerous myths about India - from the contents of the Vedas to the origins of the Aryans and invasions of India by foreigners. They have been thoroughly debunked. But it is an ongoing battle - between the scholars with little resources and establishment of centuries.

"Historically, bibliographies started in the world of mathematics! It makes sense that proofs of proven results be referred in a bibliography - rather than reproducing them again. In general, any discipline that has theories, excessive modeling and simulation, any bibliography at the end of the book should be given scant attention."

Nov 11, 2011/July, 2014

A HIST-O-MATH PREFACE

Writing the preface of a book is a crowning moment in an intellectual endeavor. The most important things about a preface are the whys, hows, and whats about the book and its author. That is what essentially follows:

How much mathematics is needed to understand this book? A little more than high school level! A history of math course can be taught with little math or with good doses of math -depending upon students' background and instructor's interest. No one is interested in the history of any one mathematics problem or topic. Yes, survey papers are common. What really needed is a mature and curious mind which can also navigate the choppy and muddy waters of math and history.

What are my credentials for writing this book? My inclination towards history of math has been organic in the sense, that in the 1980s, as I was cutting away the umbilical cord connecting me with my PhD researches, I was groping around and trying to discover my new strengths. It is a thesis that before the age of 40, there is no meeting ground between mathematical and historical thinking - the two are poles apart. Around the age of 50, I sensed a confluence of deductive thinking of mathematics and soft thinking of history gaining traction in my mind.

The ***Darts on History of Mathematics*** is a corollary of my book, ***Vectors in History*** (2012). Before writing a book on history of mathematics, I needed to establish my identity as an historian, as I approach history of mathematics as a subset of history in general. Also, if you browse standard textbooks on history of mathematics, they are stereotype and isomorphic. The contents on early history are perpetuated from one author to the other without questioning. After all, who has the time and motivation for 'revisionism'? The books are written by math professors who seldom develop a sense of history.

By including long bibliographies and lists of references at the end of the books, and lately notes at the end of each chapter, the readers are overwhelmed and the facts literally go unchallenged, particularly, if they suit a culture. Years ago, when I questioned Morris Kline on mathematics in ancient India as written in his book, *Mathematical Thought from Ancient to Modern Times* (1972), he wrote back saying that he had only quoted them from a book by **some** Indian authors. **Can a wrong proof of a theorem or a wrong solution of a problem ever be passed on by such referencing in mathematics?** Over the years, these padded bibliographies and references have been losing my respects. At least, they are not a part of anyone of my books.

What prompts the writing of this book? The darts in the title of the book precisely point towards such unverified, exaggerated and stretched out or ignored conclusions in their historical narratives. They fired me up to write these reflections on specific points while discussing particular topics. Therefore, it is not a typical textbook on history of mathematics, but can be used as supplementary material along with a regular textbook.

What is the incubation period of the book? The dates of some reflections do go back to more than 20 years, but the writing of a book was not on the horizon until relatively recently. Not all old pieces of writings were saved either.

What is new in the book? Apart from its format; in brief, it has thought provoking angles of observation and deductive conclusions on many topics, which may look ordinary or rare.

Who will benefit from the book? Any lay person with an historical bent of mind on mathematical topics stands to gain from it. Both undergraduate and graduate students in history of mathematics courses would enjoy it. All reflections are independent - they are excellent bedtime reading too.

Is there any other book similar to this in format and style? Absolutely none! That is my moment of pride. It aligns with my disinclination for writing mathematics textbook throughout my professional life of 50+ years. I do not see any challenge in writing of one. It is uncreative to shuffle problems from other textbooks and change them by epsilons and deltas! A computer will do it one day. Then, who will adopt a textbook written by a computer?

There are a total of 69 reflections and write-ups in this book. They are unevenly divided into five sections. The divisions are not sharp as dictated by the very nature of reflections. Sometimes, the overlap between topics in some reflections is so large that their variations are included in the *Scattered Matheriticles* (2010) or/and *Vectors in History* (2012). I have found some divisions better than no divisions, as was done in the first book. The first collection of 22 reflections are under the heading of **Classroom Cuts**. They were drawn from teaching in the US and overseas.

The second section, named, **Humanistic Slices** contain 15 reflections connected with the lives of mathematicians of both past and present. History wrapped around them is unraveled. The third section called **Indian Spices**, has only 8 assorted reflections dealing with mathematics of India - from ancient to the present. The fourth section is called, **Smorgasbord Bites.** As the title implies, it has an interesting mix of 17 reflections having varied shades and kinds. The creation of the last section sprang up in my mind only 2-3 months ago, when I decided to include the outlooks and experiences on history of mathematics of my friends and colleagues in mathematics. It is rightly called the **Other Perspectives**. However, it attracted only five write-ups; and they are included after some minor changes.

The book is dedicated to *"History for the Wise"*, a saying that has been registering deeper into my consciousness as the years roll by. I Googled it to know the face behind it, and found it frustrating scanning over 40 pages. I tried other search tricks, but none was helpful. My memory attributes it to Ralph W. Emerson, but it could not be tagged on him. The amazing thing is that its veracity has tested out perfectly. Really, one has no clue of history in youthful years. A life has to be lived holistically for a few decades in order for it to appreciate history.

A common feature of all my books is that they each can be read from anywhere, as they are independent in contents and topics. It fits in today's fast life styles; no one has the time to start a book from its Page Number 1 and then wade through it to the very end. Brevity is the characteristic of the twitter age - using 140 characters or less. As a consequence, abbreviations are explained again, and certain references repeated as encountered.

The practice of dating each reflection is continued so that a reader may have a full perspective of its genesis in terms of time, place and my mindset. As a young reader, I paid no attention to it, but now, this is the first thing I look for in a book. The two dates on some reflections means that a revision was done on the second date. Another continuing feature is providing partial and full blank pages for the readers to scribble their comments, as they pop up while reading it. It comes from my compulsive habit of underlining and side-lining a significant part of a sentence or paragraph. Such markings become a source of quick reference in future.

Every book ought to be a defining moment in author's life. For me, this book firmly plants my feet in history. Increasingly, I tend to look at a social scenario or problem through a lens of history. Consequently, my world of solid mathematics has shrunk. I have not volunteered to teach hard core graduate courses in PDE for nearly 20 years. Once routinely taught upper division courses in analysis, numerical analysis and matrix theory have not been taught for ten years. Even my bread and butter courses on ODE have distanced from me.

The main reason is that with the creation of **Teaching Concentration** (2002) in the MS program, my involvement has increased in the teaching of its three required courses. There are a few faculty members who can teach 'my' hard core math courses, but lately I have been the only faculty member teaching the specialized courses in this concentration. Currently, my hard core math courses are limited to Calculus, Discrete Math and Linear Algebra. At age 75, I am at peace with this shrinking choice in teaching, but it seems unlimited in terms of the books being turned out each year.

I thank UNLV and my department chairman for supporting my sabbatical leave for Spring-2014. My books are not cookie cutter types such that they can be finished within certain deadlines. Each piece overflows with passion. There is nothing mechanical or algorithmic about them. When the sabbatical leave proposal was written up in Sept, 2013, the time lines seemed different. In April, 2014, a biography of Swami Deekshanand was published 'out of queue'. So, this book was pushed into summer, where 4-6 weeks went out of my control.

Moreover, the final 1% of the book material has taken a lot of my time and energy unexpectedly. Many a time, I joked that the huge body of the elephant has come out of a gate, but its tail is stuck somewhere behind. Above all, I am far beyond any number game of the books. Each book is still written with a conviction that it would last 100 years. I don't intend to pollute the world with mundane ideas either- the mother of all kinds of pollution.

Also, I profusely thank Francis Andrew, Professor of English in Nizwa College of Applied Sciences in Oman, for providing feedback on the syntax and semantics on every reflection. With the result, along with my unabated passion for writing, I am beginning to enjoy the usage of the English language in a manner never ever done or thought before!

Finally, any comments and suggestions on the book emailed at: viabti1968@gmail.com, would be greatly appreciated.

Satish C. Bhatnagar
Oct 02, 2014

DESIGNING THE FRONT COVER

1. The material of this book is related with that of my **second** book, *Vectors in History*. Here are my broad ideas:
2. The overall color both of the front and back should be slightly light orange than it was of the covers of Book #2. However, no patches of darkness, but a few spotty shades of orange are ok.
3. In the front cover only, I want a world map in the background with boundaries of major countries, but no names at all.
4. Since no world map will fit along the width of the cover, it has to be spread out length wise. The north side of the map should be toward the spine of the book. I am enclosing free images of two maps. If you have a better map, then I would go along with that.
5. The darts in the title are to be like the darts in the game of dart board. Choose arrows of different shapes, sizes, colors, and design.
6. The darts should show up, as if they are hitting at some points of some countries in the world map – India, US, china etc.
7. The title, Darts on History of Mathematics should be on **two** lines of the upper half- **DARTS ON** in the center of one line, and **HISTORY OF MATHEMATICS** in the center of the line right below it with no line space.
8. My name, SATISH C. BHATNAGAR, should be placed near the bottom of the front cover. All letters of the title and my name should be in Font size 16 or 18.
9. The color of the title of the book and my name should be in **bright blue** instead of black or white.

More on the designing after seeing the first draft.

Oct 07, 2014

GLOSSARY AND ABBREVIATONS

Mantra is set of supposedly energized syllables in Sanskrit – potent enough to affect material changes with right repetition and enunciation.

Sutra is a cryptic and condensed description of a principle or property. An example is of 18 *sutras* of Vedic Mathematics.

Tapa is combination of penance, meditation with austerities

Vedas refer to the most ancient four Hindu scriptures, namely; Rig, Yajur, Atharva, and Saam. *Upvedas* and *Vedangas* are ancillary treatises for a systematic study of the Vedas.

Rishi is an enlightened individual in terms of his/her cultivated powers of mind developed through Yoga over a long period of time.

Guru is far more than a high school and college instructor. There is an associated element of one-one-ness, loftiness and holistic nature - bordering spirituality.

Gurukul is a kind of Hindu seminary school going back to the Vedic period.

Shrimad Bhaagvatam or Bhagvat is a holy scripture. It is often confused with Gita or Bhagvat Gita, which is a central part of a chapter in the epic of Mahabharata. Whereas, Shrimad Bhagvat was compiled by Vyas after he had finished the Mahabharata. It is a great story of life.

advaitya; non-duality - Principle of One-ness.

siddhis are the states of mind achieved after years of penance and yoga that one can materialize objects. Essentially, it is a reverse mass-energy equation. The late Satya Sai Baba of Puttapurthy had reputation to pull out of thin air jewelry items for his followers all over the world. Such a person is called ***Siddha.***

IT: Information Technology

BTI: Bhatinda or Bathinda

DLH: for Delhi, the capital of India

UNLV: University of Nevada Las Vegas

JMM: Joint Mathematics Meetings

MAA: Mathematical Association of America

AMS: American Mathematical Society

NAM: National Association of Mathematicians (Founded by Afro-Americans)

AWM: Association of Women in Mathematics

SIAM: Society of Industrial and Applied Mathematics

IMS: Indian Mathematical Society

Math: Popular abbreviation for Mathematics - used interchangeably in the book

PDE: Partial Differential Equation(s)

ODE: Ordinary Differential Equation(s)

SECTION I

CLASSROOMS CUTS

CALCULUS DEFINES CIVILIZATIONS

The study of Calculus is as important for the total development of the young minds, as is the study of Shakespeare in literature, Socrates and Plato in philosophy and critical thinking, and of Aristotle in political science and history. Calculus sits on the concept of **Limit** that crystallized over a period of two thousand years. **In the entire gamut of human thought, there is no concept as profound, and yet as practical, as that of Limit in Calculus.**

Calculus is truly a defining landmark in the history of Western Civilization. Surveying contemporary civilizations, since the 11th century, Europe was then in the dark ages and America undiscovered. The Islamic empires were raging in the Middle East, North Africa and Eastern Europe. The Mongols, ruling over China and stretching into Europe, formed the greatest empire in history. Travelogues, like that of Marco Polo, describe the glitter of Mongol royal courts in China. Since then, it has inspired generations of Europeans to seek their fortunes in Indo-China.

In the 17th century, the Mughals built one of the most prosperous empires in North India. But mathematics and science did not make considerable headway despite the knowledge of Hindu Numerals and decimal system. However, their power was unleashed when the Hindu Numerals replaced the Roman Numerals in Italy in the 14th century. It was the beginning of European renaissance.

The discovery of Calculus was a quantum leap out of the union of deductive reasoning of mathematics and experimental demonstrability of modern science. It actually burst into the firmament when Newton came upon the scene in the 17th century. He remains the greatest mathematician and the greatest physicist of all time. Around this period, the European Civilization caught up with others. Calculus is at the heart of progress in sciences, and it eventually became a spark plug in the engine of industrialization in Europe.

Since the rest of the world had no clue of Calculus, it continued to lag behind Europe in mathematical discoveries and scientific inventions. During the 18th and 19th centuries, it naturally propelled Western Europe to explore the world that also resulted in the colonization of Asia, Africa, America, Middle East, and Far East. Though by the middle of the 20th century the decolonization had started, the West still continues to forge ahead. **The power unleashed by Calculus in 17th century transformed into nuclear power in the 20th century!**

The magic of calculus lies in the fact that it lets one have the glimpse of Infinity, and what is potentially infinite. The limiting processes are inherent in human experiences. For instance, it is in a 'limiting' process that average velocity becomes instantaneous, and a line joining two points on a curve turns into a tangent at a point. Many ancient paradoxes, like that of **Zeno,** were laid to rest once the concept of **Limit** was grasped. Division by zero is a mathematical no-no in high schools, but I characterize Calculus as a story of Division by Zero, or a Story of Infinity!

Calculus is the ultimate experience of a rational mind. Above all, **a grasp of Infinity through Calculus brings one closest to an understanding of the Infinite Manifestation** - call it a god, or the God! So who would like to deprive oneself or others of its taste and beauty? As educators, we owe every student some exposure to Calculus. It is a giant step of mankind in its intellectual evolution.

June 09, 2003/June, 2014

[PS: This is one of my all-time favorite reflections. It is widely circulated amongst students and mathematically inclined readers of my reflections. Also, its variations have appeared in a mathematical journal and in my first book, *Scattered Matherticles*; *Mathematical Reflections*, Volume I (2010)]

ON HISTORY OF MATHEMATICS COURSE

A history of mathematics course is questioned for its rightful place in mathematics departments. It is quite challenging to teach it. On the other hand, if mathematics departments won't offer it, then who else would care about it. There are all kinds of history courses in the US institutions. The esoteric ones are like, history of religions, history of philosophy, history of languages, history of news media, history of women studies, besides the popular ones - history of various arts, firearms, fashion etc. UNLV's History Department offers a course on History of Science. What kind of slant would it have, if the instructor never had even calculus and science courses?

For the last 25 years, general interest in history of mathematics has been steadily growing in the US colleges. History of Mathematics is the only topic in which the sessions are organized by all mathematics organizations viz. AMS, MAA, SIAM, NAM, and AWM. History of mathematics sessions often attract the largest number of audience. Moreover, I have seen the top notch mathematicians sitting in the audience. Its popularity is like that of shopping in WAL-MART - you get both – all items under one roof and value! Also, the MAA has a special interest group in **History of Mathematics**.

There are a few caveats about such a course. One does not need a course on history of mathematics in order to learn mathematics at any level. However, interjection of historical remarks in the context of a proof of a theorem, solution of a problem, and idiosyncrasies of a mathematician always adds spice to the instruction and topics. Students' attention span in math lectures being short, historical touch is always appreciated at the undergraduate level.

There is a famous line of George Santayana: **Those who don't remember history are condemned to repeat it**. It is very true in the world of politicians, but it has indirect bearing on mathematics. For example, history of mathematics is solidly tied with the exodus and extermination of the Jewish mathematicians from Nazi Germany during the 1930s and

40s, and the emigration of Russian Jew mathematicians in the 1990s after the collapse of the USSR.

History brings linearity of timeline which is significant for measuring progress. On a personal note, while studying in Bathinda, a hinterland town in Punjab, objective history of India was not taught, even after India was free in 1947. History of mathematics was never in any realm of thought. The first book on *History of Indian Mathematics* was published by Datta and Singh in 1962, bringing out India's mathematical heritage.

UNLV Math Dept has two history of mathematics courses; one (MATH 314) for the undergraduates majoring in high school teaching, the second (MAT 714) for the MS students in the *Teaching Concentration*, one of the four concentrations in the MS program. Though I never taught MATH 314, but I am excited about teaching MAT 714.

My interest in history of math is a corollary of my continued interest in history in general. Besides, presenting papers in MAA history sessions, in 1985, I offered the first course on *History of Modern Mathematics*. For me a personal challenge is to relate history of mathematics with social and political conditions conducive for it, and its disciplinary connections and parallels.

Jan 04, 2007/June 2014

COMMENTS

Dear Professor Bhatnagar, It is good with some striking observations, only it could be rendered mathematically better. In the sense, let us stress more on history of mathematics taking clues from general history, and not involving UNLV, or else it becomes more of a personal reflection, as you have been engaged in.....Let us leave this to what others have experienced and quote them appropriately as you have done at some places. .....**BSY**

Dr. Bhatnagar- Very brief and to the point. I see you are gearing up for your course. Have you selected a text? What were your papers on that you presented to MAA? The more I research my professional paper, the more I come to see the importance is for the math teacher to have an understanding of the history of mathematics more so than the student. I am still going forward in my paper as I was. Recently, my students and I discussed Pascal. I think they enjoy the information and readings even if it does not have an effect on their grade. We will see however when I put it all together. Thank you, **Cassidy**

MY PERCEPTION OF HISTORY IN GENERAL

History, at any degree of depth, does not belong to the historians in the academe alone. At any given time and place, there are history makers, history readers, and history keepers. In a limiting case, each individual knowingly and unknowingly combines all the three facets of history at different points in life.

While undertaking to teach a graduate course on *History of Mathematics* (MAT 714), I questioned my credentials. Being in the business of mathematics teaching and research, my mathematics background stands on solid footing. For justifying my history credentials, besides having taught a lower division course on history of mathematics, I love to interject historical anecdotes in every mathematics course that I teach.

The Hindi equivalent of 'history' is *itihas.* It literally means story of the olden times. The word history, having roots in Latin and Greek, comes from *historia*, meaning inquiry, to know etc. In common parlance, history is a chronological record of significant events. (Mathematically speaking, every nice non-linear function at a point can be 'locally' linearized.) The word 'significant' is very subjective.

History does not occupy the same prominent place in every culture. The Hindus, with their faith in reincarnation and the *Law of Karma*, have a cyclic view of life and death. Hence, the chorological linearity is absent in Hindu way of life. History is not discerned in any ancient treatises of India. Even the big time Hindu kings never chronicled their reigns as compared with the Muslim kings who commissioned artists and court chroniclers for their *shahnamas* – means, royal chronology in paintings and words.

In contrast with the Hindus, the Sikhs have taken their 500-year old short history to newer heights. Prominent gurdwaras (Sikh temples) have permanent exhibits on the martial side of Sikh history. During *ardaas* (concluding Sikh prayer) homage, to the men and women who sacrificed their lives for Sikhism, is an integral part of invocation. The first series of history books, *The High Road of Sikh History* prescribed during my junior high schools years were written soon after India's independence in 1947. By the 1960s, HR Gupta and Ganda Singh were well-known Sikh historians of the 1960s.

By and large, history in Asian countries continues to be all about the events affecting the kings, presidents and prime ministers. On the other extreme, the American culture, amalgamated and transplanted from the European cultures, want lineage and continuity of ideas and persons in newly discovered lands. They are obsessed with 'history making' - from erecting the historical markers on the roadsides and hiking trails to the museums of natural history etc. Anthropology and archeology (space, land and marine) are corollaries of this mind set, as these disciplines have no roots in any other cultures; past or modern.

Until recently, the history of India existed mostly in the form of diaries written by the British officers posted in India and by Indian scholars under the British influence. They have perpetuated distorted facts on India. It challenged me, while I was still in junior high school. Because of the curricular restrictions, history and mathematics could not be studied together in college. However, I have continued to explore history on my own for the rest of my life.

During the last ten years, several articles and *Reflections* on different aspects of history have been written up. The preciseness in thought, due to my mathematics background, gives a unique dimension to my discovery and examination of historical facts, their interpretation, and relevant connections with the present.

Now entering into my 70s, I tend to understand why history is considered for the wise, and how the study of history makes one wiser. On Day Number One, in a diagnostic history quiz, I challenged my students to bring any piece of record on their great grandparents. I don't know even the name of my great grandparents! But the record keepers in Hindu holy places have been maintaining the genealogical records of the pilgrims for centuries. My father had told me that a full-fledged genealogy of our family exists, and it goes back to the year 1560. Somehow, it was never passed on to me.

Last week, one student brought excerpts of his great grandmother's life written 80 years ago. It was a joy to go through them. Another student brought meticulous genealogical notebooks that her great grandmother researched and traced back to the 16th century. She sought its retrieval assistance from the genealogical archives of the Mormon Church in Salt Lake City, Utah.

I really want my students to have an unforgettable hands-on experience on an aspect history of the last 50 years. A span of 50 years, in the life of an institution, is a speck in a time line as compared with 'recorded' history going back to 2500 years. Students, doing research, immediately click on the Google searches, or run to a library. To be able to find a source of information is only searching, it is not doing basic historical research!

UNLV was founded in 1957. Its one-year long celebrations will begin in fall semester. To make an 'historic' contribution, I decided to discover the history of our Mathematical Sciences Department from every conceivable aspect; faculty, students, programs, curriculum etc. The project has been divided into ten smaller projects and assigned to the ten students registered in the course according to their interest and background.

History and mathematics have one thing in common. **It is easy to pose a good question, but very difficult to find an answer**. The contrast is whereas there is little disagreement amongst mathematicians over a proof of a mathematical assertion, but if two historians agree on a fact, then one is faking it, or taking a free ride. Finally, discoveries and inventions define the homosapiens from the rest of species.

Feb 04, 2007/May 2014

PS: This reflection is essentially included in my book, *Vectors in History* (2012) which was written as a precursor to this book.

COMMENTS

Satish, I really enjoyed your take on history. Perhaps a tendency to want to record events for future generations is one thing that differentiates (to use a mathematical term) humans from animals. Of course animals, in a sense, record history in their genes. **Dave Emerson**

PERSONAL REMARKS

HANDS-ON HISTORY
MAT 714 (HISTORY OF MATHEMATICS)
INDIVIDUAL RESEARCH PROJECTS

The whole idea is to have a first hands-on experience of doing research in the historical context of mathematics. It is a good starting point as you are not going back in time beyond 50 years. This exercise shall open your mind to the challenges of finding general historical facts, later in life.

As UNLV begins celebrating its 50th anniversary from fall 2007, these projects may provide different insights how the entire enterprise of mathematics was at UNLV 50 years ago in terms of its faculty, students, curriculum etc.

Here are my ideas on it. In case, you think of any thing else, let me know and I will include it. The list is not in any order. In case, you discover any tidbits, anecdotes, do include them, like oral history. Identifying each source brings integrity to the data. For history, nothing should be assumed as trivial.

1. **Math Faculty**: names, qualifications and institutional affiliations – BS/ MS/PhD.
    Chairmen: Acting/regular. What happened in the Department during their leadership positions? Include folklore items like, David Ashley as the 8th president of UNLV. It takes away the dry edge of history
2. **Calculus I**: Course descriptions, textbooks and their authors, instructors, instruction formats.
3. **College Algebra**: Course descriptions, textbooks, their authors, and instructors; instruction formats. Currently, there is Precalculus I that is 'equivalent' to College Algebra. Find out if and when these two courses were created; one for students in colleges of Business and Hotel, and the other for colleges of Sciences and Engineering.
4. **Trigonometry and Analytic Geometry**: course descriptions, textbooks and their authors and instructors, formats. It would be interesting to find out whether there was ever a course on trigonometry, or it was always a part of college algebra. Currently,

there is Precalculus II that alone covers trigonometry. Find out when trigonometry turned into Precalculus II.

5. **Precalculus I & II offered as one course**: The origin of this course name, Precalculus, and its evolution at UNLV, and/or perhaps across the nation.

6. **Intermediate Algebra**: Being the first remedial course at UNLV, find its first offering, descriptions, textbooks and their authors and instructors, formats (large, small sections or individualized section).

7. **Elementary and Pre-Algebra**; Being second and third remedial courses in the Nevada, Higher Education System, find their first offerings; course descriptions, textbooks, their authors and instructors, formats.

8. **Undergraduate Math Majors**: History of BS and BA programs, the number and names of math majors; Undergraduate mathematics coordinators, if any. Having addresses of math majors may create a pool for Math Dept alumni.

9. **Math Graduates**: History of the inception of the master's and PhD programs, Graduate Coordinators; the number and names of math majors. Having their contact information including addresses may create a pool for Math Dept alumni.

10. **Women in Mathematics at UNLV** in terms of faculty, math majors both undergraduates and graduates.

Please send me a couple of lines of your progress by the end of a week or two, or whenever feeling your progress stalled. Keep a detailed log of your research time and efforts. I 'assure' you it will be an unforgettable experience at the individual and collective levels.

The undergraduate seniors may like to enter their research projects in an MAA (Mathematical Association of America) History of Mathematics competition, if finished by the deadline of March 31.

Feb 01, 2007/Jan 27, 2010/May, 2014

HISTORY OF/IN MATHEMATICS

This semester, teaching a graduate course in *History of Mathematics* (MAT 714) has been a wonderful experience. Partly - because, it resonates with my interest in history at large. Often I bring general history into mathematics and connect the development of mathematics with that of sciences, political systems, and even with major religions. Roger Cooke's *History of Mathematics* is used as a textbook for keeping some structure of the course. But the class discussion and other materials dominate the contents.

At the beginning of the semester, I included 2-3 class tests depending upon the pace and coverage of the material. Though I was really not clear on the type of test questions, but felt it necessary to frame them for my own record of class discussions. Of course, I eased students' concern by de-emphasizing the dates, a bane of most history courses in schools and colleges.

It took me a full day to think, write and choose 12 questions for a 75-minute **closed** book test. I told the students that for them it would be like finishing a gourmet dinner in an hour that the chef had spent half a day in planning and preparation! But the fun began when I started grading the answers. Unlike in a math test, **there is no wrong answer in history**! The maturity, professional experience, preparation, and course involvement of the students were certainly reflected in their answers. Normally test grading is a pain for the instructor, but for the first time it was fun of its own kind!

On a serious note, I thought why a history of math course is not required for math majors? Any one for BA/BS from most US universities must take a course on the US History and/or political system. Those majoring in the areas of arts, business, humanities, law, and social sciences, eventually do in-depth study of their histories. Somehow, the disciplinic history is not included in sciences and mathematics except for those majoring in school and college teaching. At UNLV, MAT 714 is not counted for credits towards MS with concentrations in pure math, applied math, and statistics.

Being a hard core mathematician by training and practice, I fully understand the dilution factor of the curriculum with the inclusion of history of mathematics courses, since it has little prescribed mathematics content. **But it should be encouraged as an elective beyond the minimum credit requirements, then**. Its life time payoff is huge.

However, the main point is changing the public perception of mathematics and its practitioners by having such a course accessible to the students at large. It would make mathematics degree programs humanistic. The gap between mathematics and other disciplines may be narrowed and bridges built over. This is an era of public relations – political lobbyists, accountability, assessment, multi-disciplinary, diversity, image, and applications. Mathematics being a linchpin of natural and social sciences, it is time that mathematics departments engage in self-examination. A good course on history of mathematics can go a long way in serving this purpose.

Any way, here are my 12 questions on **Test #1** that I think are good enough at least for the back burners of mathematics graduate students and instructors:

1. Give an **example** of a society or nation where mathematics has flourished in **isolation** from other **disciplines**.
2. Amongst the **ancient** nations or civilizations of India, Egypt, China, Greece, give one example each of their mathematical prowess of the **past**. (Assume that the national boundaries change every 50 years.)
3. Calculus concepts are classified either **local** or **global**. For example, **Limit** is local being defined at a point. **Integration** is global being defined over an interval. In life, **caring** for the family, nursing the sick, serving in restaurants, working in front offices, etc. - women have a dominant presence. Despite the fact that women excel in 100-level math courses, **they don't pursue mathematics for being less challenging**. Does the motherhood provide them a higher experience in creativity and caring; your comments?
4. Explain the reasons for the **differences** in the present state of mathematics between Mexico, south of the US, and Canada north of the US.

5. Mathematics is the **index or measure** of the prosperity of a nation. Justify or contradict this statement with examples.

6. Based upon your hands-on research experience in individual class projects, what lessons have you drawn about the **historic facts and research in history**?

7. One of the themes during class discussion is the role of the **organized religions** in science and mathematics. In the context of the major religions, explain this **correlation** with examples from the past and present.

8. In the **Preface**, Roger Cooke remarked, "….writing the history of mathematics is a **nearly impossible task**." Comment upon it. Name any **two other** books on the history of mathematics.

9. Whereas, the truth-value of a mathematical statement/theorem **does not change** with time, it is not the case with the historic facts/truths in general. Explain with examples.

10. What is the **strength** of the Hindu Numeral system over other number systems?

11. What is **unique** about Greek mathematics over contemporary mathematics known in other cultures?

12. What has been the most **useful** feature of this course so far?

Mar 18, 2007

CONDUCTING HISTORY PROJECTS

I am already beginning to feel like a conductor of a musical symphony where first, the instruments need to be tuned before playing them. The next phase is the mastery of individual compositions, and finally putting music and musicians in a perfect stream. These history projects may appear daunting, but I remain optimistic. One of you e-mailed update, and another solicited information in the class. Putting all your 'raw reporting' together, here are my thoughts to assist you further:

1. You and your project are not like in a boxing bout where you try to knock the opponent out in one big punch in the very first round. It does happen in boxing, but more often the fight goes to the last round, or historically, till the other fighter can't get up.

   A lesson from your experience, so far, is that there in no one place on-line or off-line where you can go and make prints or copy the material down. If there were such a place, then it would not be called 'your research'? It is a search then.

   Paradoxically, research looked at as 're-search' means **searching something that was 'lost'**. It makes sense in arts, humanities and social sciences, **but in the sciences, research means to find that has never been invented or discovered before**. But you need not take any sides on it; be a music rapper and mix the two.

2. Any research is collaborative. It is also sharing what may be useful to the others. For example, if one working on math faculty also finds curricular information on the side, then it should be passed on to that classmate. Immediately, use text messaging, phone all, or e-mails. But never forget to pay back and give credit to that person.

3. By now you must have realized the importance of resource persons who have been around the department for a long time. **Getting them to help you requires a communication skill**. Why a person should help you? Be prepared to 'pay' for it. You know there is a book on this subject that librarian, JD Kotula had pointed out.

Do some homework before approaching the resource persons; your questions are well written up and delivered in advance -giving the person time to think. No one has the data on finger tips. Of course, some may be preoccupied, or say a blunt No. Then, you don't feel bad. I have spoken with Paul Aizley about the projects and in fact gave him a copy. So, you go/call/e-mail him with full confidence; better meet him in a small team.

Malcolm Graham, the first chair, Lloyd Nietling/Acting chair, Lewis Simonoff/Acting are still alive! The 1960s graduate Students like John Green are around too. Research in history is a detective work - one clue leads to the other till a puzzle begins to unravel its mystery.

I look forward to interesting updates next week. Keep it in mind, while you enjoy the world's Greatest Symphony, the American Super Bowl!!!

Feb 06, 2007

A SAMPLE LETTER FOR HELPING STUDENTS

I am taking time to write you individually for seeking your assistance in re-creating the history of the Department of Mathematical Sciences since UNLV was established in 1957. It coincides with 50th anniversary celebrations starting in fall 2007. This project is divided into ten 'smaller' projects as there are ten students registered in the course, *History of Mathematics* (MAT 714) that I am teaching this spring. It will provide the students a hands-on experience on a piece of history. The appendix below contains some details on each project and corresponding names of the students.

For investigating into projects - dealing with faculty, students, programs, courses, no record is to be ruled out, no matter how trivial it may appear at the outset. It may provide some clue to the next source. We all cherish and have files and folders on these matters. Take time to look into your archives as in some cases after retirement in particular, their contents may have been forgotten.

The path of least disturbance in your daily routine is that you let me know whatever material you have by next Monday (02/12/07). There are two options for accessing it by the student researchers. One, you deliver or bring the records to the Math office. Two, let me know what days and times are convenient for a team or 2-3 students to visit you.

As you know the oral history is integral to historical events. If you wish to talk with 2-3 students at your convenience, then it can be arranged. I know recalling old facts and then writing them out is too much. In history, I believe, one clue can lead to the other –like, the deductive steps required in proving a theorem.

I know you may have trashed some of the records after a lapse of 15-20 years as most of us do it. But I must repeat that any little help may be great in putting a big picture together. It is like assembling pieces in a zigzag puzzle. I assure you of confidentiality, if that may be of concern at places. Your cooperation will be duly acknowledged and appreciated.

With personal regards and thanks.

Feb, 2007

PUSH - PROJECT - PROGRESS

History is created every minute by every individual, knowingly or unknowingly. Likewise, the process of unearthing any old piece of history is a sum total of all tiny efforts in every resource taking time and energy. It is time for every one to pitch in extra effort in his/her individual project and join with someone else's too.

Any data base, or for that reason any bank account, is as good as we put into it. The more we invest, the more useful it would be for us and others. We are making progress in each project, though at varying speeds. I am confident that the final product will make us all proud. These remarks are prompted after gleaning your updates last Tuesday.

Over this All-Star NBA weekend, I followed up with calls to Emeriti professors, Malcolm Graham (255-0619), Harold Bowman (732-1901), Lloyd Nietling, Tom Schaffter, and Looy Simonoff. They are all receptive! Don't hesitate to e-mail your specific questions or put them in their mail boxes, or call them for a meeting/conference in a team of 2-3 in the math conference room/lounge, or *Book N Beans* coffee place.

There are two more solid resources waiting to be tapped. PTI (part time instructor) John Green told me only yesterday that nobody had approached him yet! He has been around since the late 1960s. He has a good memory besides holding onto a lot of old material. The other PTI is Tony Blalock who finished his MS in late 1970s. Both of them have mailboxes and offices in the Department. A lesson of historical research is the importance of oral history, and hence of meeting people who are sitting on a pile of information and having no idea of its value.

Of course, some of you are mining on to professors Aizley, Verma and Miel. But make sure your individual inquiries are taken care too. The more interest you show, the more likely it is that you will elicit more information from them. Having specific questions on each project and apprising them by e-mails/mailboxes will prompt them in those directions. Let a set of your questions be a part of your efforts for the next individual project update.

As the month of Feb draws to a close, let us work for its tangible shape. Oh yes, I took care of the online form needed to get information from UNLV. Keep me posted about it.

Feb, 2007

PERSONAL REMARKS

RATIONALE FOR HISTORY OF MATH COURSE

"History is for the wise", said the great 19th century American thinker, Ralph Emerson supposedly. History of mathematics, like general history, is seldom appreciated during early stages of studentship or professional careers; one has to attain a degree of maturity and experience in life. In mathematics, one needs to be free from the professional pressures.

During the New Math Movement of the 1960s, the emphasis on concepts over computation in schools and colleges turned mathematics for the 'elite' students. The textbooks, filled with mathematics rigor even in elementary courses, were void of any history of topics or mathematicians. Thus, popularity of mathematics nosedived. The reformation started in the 1980s when computer science exploded, it pulled the kids away from mathematics. Though the overheated IT market has cooled off, but mathematics is not a popular academic destination for the undergraduates.

A required upper division history of math course brings curricular diversity. Incidentally, diversity is a buzz word for the hiring and recruitment of minority. This will also spice up traditional mathematics curriculum. The only prerequisite for the course is a senior standing so that the students are already exposed to all the major topics in mathematics.

The idea came off this semester while teaching a graduate course in *History of Mathematics* (MAT 714) required for students doing MS with Teaching Concentration. It does not count for credits for concentrations in Pure Mathematics, Applied Mathematics and Statistics. My experience is that there is tremendous flexibility in the material by weaving several topics into one strand or several mathematicians around one topic. History brings out national prides and roles of civilizations. Above all, mathematics becomes humanistic!

Such a course can be **capstone** course of a program. It can be used to satisfy **technical writing** requirement, if any, or used as an **assessment** tool of a program. Assessment is another buzz word in academia. Such a course raises total awareness of the students as mathematics is observed from a wider stage. A question may be raised: Does it dilute the mathematics contents? It being one of the at least 12 courses required to major in mathematics, my answer is No.

Mathematicians rarely think in terms of the **PR** values of their courses since they are carved in stones. This course, being prone to innovation, can have a long term impact on the minds of the students even after graduation. Above all, if math faculty won't teach the history of their bread and butter, then who else would care about it? If the attendance at the sessions on History of Mathematics at annual meetings and summer fests is any index of popularity, then such a course should be in the books of every math department.

April 30, 2007

A MATHEMATICAL EXCITEMENT

After 60+ years of lifetime, it is not easy to get excited enough whether an encounter is physical, or mental; forget the fuzzy spiritual! The signals that the mind sends to the brain are generally ignored by the body. Aging begins to calcify the nerves and arteries. By the same token, ironically, one starts forgetting moments of real excitement from the past!

However, a couple of days ago, I told a colleague during a hallway conversation that I never felt so much enthused as about teaching a course on *History of Mathematics* (MAT 714). Last time, I taught history of mathematics was a one-credit experimental course (MAT 401X). It was offered during a 3-week winter/mini term, in 1985. **History, in contrast with mathematics, is not for the young.** That is why every one hates history in high schools, and a few like it in colleges. Its appreciation begins when one hits into any social aspect of public life, and then it lasts longer than of mathematics. Eventually, some mathematicians do become historians of mathematics.

Two curricular thoughts struck me while teaching this course. Though the experimental offering was very successful, but it was not proposed as a regular course. Since 1980, history of mathematics has been very popular amongst the researchers, and educators in the US high schools and colleges. All major mathematics organizations, AMS, MAA, NAM, AWM, and SIAM sponsor history of mathematics sessions during their regional and annual Joint Meetings.

At UNLV, there are two courses on history of mathematics; one at the undergraduate level (MATH 314) and the other at the graduate level. Both the courses are for the students pursuing mathematics teaching. **It is time to require this history of mathematics course for all math seniors.** Besides its long term **public relation value**, there are **five other reasons**. This course can be used as to **assess BA/BS programs**, to satisfy the **technical writing, capstone course, multi-cultural/diversity,** and **international** requirements found in the general education core of most US universities. Above all, if mathematicians won't teach the history of their discipline, then who else would even care about it?

Graduate students often take filler courses towards the end of their program. I recall taking physical conditioning and computer science courses while working on my doctoral dissertation at Indiana University. At UNLV, history of math is a requirement in the MS program with teaching concentration, but it should be strongly recommended for students in other concentrations too. The presentations in both the courses are not free of mathematics contents, but the emphasis varies. Its greater value lies in the evolutionary understanding of mathematical problems, mathematicians, and intellectual culture of the era.

Rogers Cooke's textbook provided reading material, but it was supplemented by numerous handouts. Class discussions were very open ended. Guest lectures from the library, history and mechanical engineering provided lot of stimuli and different perspectives.

The highlights of MAT 714 were **nine** hands-on research projects. Collectively, they have created the first history of the Mathematical Sciences Department, as UNLV prepares to celebrate its 50 years in fall 2007. The time and efforts that went into the research was significant - 10-13 weeks. It was a very innovative part of the course.

Just to give an idea of the nature of material touched upon, here are 20 problems from the 2-hour closed book exam given last week. It carries 15 % of the grade:

1. The shapes and mathematical properties (like Pythagorean Theorem) of objects in two or three dimensions, as focused in geometry, are intuitive to **human intelligence,** whether expressed in the art forms or mathematical ones. Give your comments.
2. Write a **brief** history of π.
3. What are the **main sources, cited in the textbook,** of ancient (more than 1000 years old) **Chinese** and **Hindu** Mathematics?
4. The world is sharply divided between haves and have-nots when it comes to **mathematics**. Elaborate on it.
5. Based upon your research experience in a project connected with the history of Department of Mathematical Sciences, what are the **three** major obstacles you encountered in collecting data and

getting information particularly from the Department, and the University at large?

6. Of the 25 great Greek mathematicians (BCE), some are known for philosophy, name only **three** who are **primarily known for mathematics**.

7. Give a geometric explanation of the classical problem of **squaring a rectangle** with sides a and b. (Constructing a square whose area is equal to the product of a and b.)

8. What is the **Method of Exhaustion** attributed to Eudoxus (4th BCE) and its connection with Calculus.

9. Of all the great civilizations that are at least 500 years old, why art, science, literature, and mathematics of ancient Greeks alone have survived with **amazing** abundance and authenticity?

10. Comment on an observation: whereas, science and mathematics contribute in the building of great empires, arts and humanities accelerate their decline.

11. State Euclid's **Parallel Postulate** and comment on its role in the creation of non-Euclidean Geometry.

12. The controversies between **creationists** and **evolutionist** are also due to institutional splits (like separate Bible and theology colleges). The study of science, mathematics and religion is no longer under one roof as was the case during the 16th through 19th century that produced minds great in science/math and religion. Your comments.

13. Name **three** Islamic and **three** Hindu mathematicians who flourished before the 13th century.

14. Briefly comment on Rogers Cooke's remark, "**Algebra suffers from motivational problem**."

15. The word algebra is a derivative of the Arabic word, '*al-jabr*'. Does it mean that the Arabs were the first to invent algebra? State your perspective on it.

16. Briefly comment on Rogers Cooke's remark, "**The great watershed in the history of mathematics is the invention of the calculus**."

17. Simply explain the **Petersburg Paradox** in probability.

18. What is the story behind the **Simpson's Paradox** in statistical inference?

19. Roger Cooke: "**Logic has been an important part of western mathematics since the time of Plato.**" I think **mathematical logic** is integral to **western society**. Add your comments.
20. How would you look back at this course? In particular, did it enhance any **academic** strength?

May 13, 2007

COMMENTS

Always enjoyable reflection, I hope that you get some response as to changing the curricular requirement. I do think that the MAT 714 course should stay as a requirement for the MS Math Teaching Concentration (students always appreciate a little historical comment).

Also, I had some students that I had selected from my school receive the Benjamin Banneker award (I could not attend due to my own finals). Coincidentally, I was asking one of them about it and he told me you were one of the guest speakers!! Sounds like you're putting all your toastmaster's talent to work! Cheers, **Aaron**

Dr. B- Thanks for adding me to your mailing list. As I read your thoughts on the excitement brought on by teaching history of math, it occurred to me that we all need a bit of variety in our teaching. While teaching at Garside Junior High in the 90s, I typically taught one class three times and a different class twice each day, and I deliberately avoided following my lesson plan the same way in each class. I think teachers need to do this to avoid becoming narrower and narrower.

Years ago, we used to talk about the difference between a teacher with twenty years of varied experience, and a teacher who has had the same year of experience twenty times. I'd rather work with the former type, and I certainly don't want to be the latter type.

I'm reminded of a colleague from forty years ago at University of Northern Iowa. He taught their version of Math 122/123, the classes I usually teach. He had taught them so often that it became automatic, something like turning on the same video and running it for each section. He grew tired of it, and one semester made a deal with his department chair and his dean---no one else knew of the plan. He taught one section his usual way, but he taught the other in a deliberately bumbling fashion. He'd make obvious mistakes, or he'd act like he forgot what he'd just done or what he planned to do next. Now and then he'd use the wrong term or "forget" the meaning of a term. When the semester ended, the SEOCS data showed that one class thought he was the best Prof on campus, and the other thought he should have been fired long ago. The exam grades were inversely related, however. The class that thought they had a bumbling fool for a Prof did much better. His name was Bud Trimble, and he was, at that time, the primary author of K - 12 math texts for Scott-Foreman publishers. **Owen Nelson**

A PUDDING PROOF!

The final exam of *History of Mathematics* (MAT 714) has been over for two weeks ago now. Looking back at a finished task provides new insights. One of the challenges of the final exam was finding the right questions and then phrasing them. It is so different from typical material for a calculus course. I took the opportunity of writing this exam as time to look back at the material covered from the textbook and the one I provided and emphasized in class discussions. It was an educative exercise. This exam turned out like a gourmet meal that a master chef spends an entire day in planning and preparing, but the guests wash it down in an hour!

The 20th and the last question in this 2-hour exam was: *How would you look back at this course? In particular, did it enhance any academic strength?* This question was pertinent, as all the nine students were either currently teaching, or are going to be teaching. Their motivation showed up in their weekly projects and class discussion. Well, here are all the nine responses:

1. "….has deepened my appreciation for mathematics and its history. It has broadened my views of different cultures and has shown similarities of these cultures through mathematics. It made mathematics more humanistic and I wish to incorporate them in my teaching."
2. "….was insightful in that it gave me an opportunity to understand the difficulties of doing historical research…..It definitely enhanced my knowledge of the history of mathematics….I apply this information in my classroom weekly."
3. "Our critical thinking was pushed to its absolute limits, while broadening our horizon. …made me want to integrate so much wonderful information from the past into my lecture…"
4. "…made me research more than I have for any class at UNLV so far…….I did enjoy my time in the class and enlightening discussions."
5. "There was a lot that I learnt in this course……I enjoyed renewing my research skill and researching topics of interest…"

6. "This is a unique course...I have a better appreciation for historians. ... learnt some of the skills needed to do true research.... Know more about the development of mathematics in other countries. Knowing about math backgrounds of other civilizations makes men more relatable."

7. "...it developed my analysis of this development of history and mathematics. ...became more aware of how mathematics was developed through the ages and when....."

8. ".......It gave me a better understanding of how some branches of mathematics such as calculus came to be and the order in which mathematics was developed."

9. "...helped me understand the human side of mathematics. Naturally I knew that mathematics had existed, but to actually consider their work from a human side was very revealing. Additionally, the historical/research aspect of the course made me question my own understanding as well as the claims of others. I am much more attuned to historical claims always wondering what other part of the story is missing."

Ten students had registered into the course. One student who stopped coming after 2 weeks due to medical reasons had 1962-PhD in statistics from Columbia, an Ivy League university. Two were graduating seniors. Another student had an MS from IU and teaches in CSN; two high school teachers, and the remaining four, UNLV graduate assistants.

May 24, 2007/June, 2014

CHALLENGES OF FINDING FACTS

Innovation in historical research comes either from the new facts uncovered, or asking old (new) questions in a new (old) context, and then re-examine the existing material. History has been my hobby, but teaching *History of Mathematics* (MAT 714) has opened new vistas. It was during the annual Joint Mathematics Meeting that I heard a session speaker suggesting research topics as a part of history of mathematics instruction. The spring offering of this course coincided with the upcoming 50th anniversary of UNLV.

Taking into consideration students' mathematical interest, the topics were accordingly assigned to all the ten students right in the second week. Again, I remembered the speaker's comment that a topic may appear simple, but the research into it would make it deeper. That turned out absolutely right in all the cases. The topics were curricular and personnel - dealing with remedial courses, college algebra, precalculus, calculus, undergraduates and graduates, female students and faculty, and leadership of the Department since 1957. The research outlines were open ended. Neither I could foresee them, nor did I want the students restricted, if they enjoyed pursuing a research trail.

The weekly updates kept each other on the toes. Having been at UNLV since 1974, I provided some clues and directions to keep the students focused when they felt stone walled. Since all the projects were to collectively create a history of the Department, it was essential to see each project finished. Initially, 6-8 weeks appeared sufficient, but a few went on for 10-13 weeks.

At times, historic facts are elusive like the existence theorems in mathematics that provide no algorithms for finding the nuggets. We all know that the students have graduated, faculty were hired and textbooks used; but with available resources, limited time frame, and the stakes, the complete answers defy the search. Some holes and gaps may be plugged by the next group of students taking this course. UNLV being a young institution, there are no apparatuses for creating its historic archives.

At the end, I invited the students to specifically comment on institutional obstacles encountered during their researches. In summary, the most common gripes were bureaucratic bottlenecks; partly because the staff was never approached with such inquiries. However, a librarian provided useful tips on doing historical research and conducting interviews etc. There is no one place for all the university catalogs. '**Protected Personal Information**' was a pet peeve. The departments need to develop their archives - both classified and de-classified. The material in the **Special Collection** of UNLV libraries is not organized and its opening hours are limited. Alumni, Registrar, Graduate offices refused access. I filled out an online form for an office; still the data was not enough.

An interesting part of this research was that the students realized that you can't simply Google research topics from the comfort of home or office. **They learnt the value of leg work, communication skills with different offices, and interviews with chairs and faculties**. Still, no one approached the five emeriti faculty including the first math faculty and Dept Chair, Malcolm Graham. He joined UNLV in 1956 when it was an extension of University of Nevada Reno!

May 26, 2007/June, 2014

COMMENTS

If you could write a concrete report of the actual work done by the students during the 10-13 weeks with statistical support (not simply words) followed by important conclusions and suggestions for the future work to be pursued, I may get it examined for publication in Ganita Bharati for the 'Classroom Notes' section. Mention of handicaps and shortcomings faced would be added information. The whole article should appear as an actual report of a project undertaken, rather than simply a reflection. Thanks. **BSY**

I wrote: I can add an appendix containing some detail on each project, but I need to know the audience of the *Ganita Bharati*. How many and who subscribe and read it? My mathematical Reflections are read by at least 100 persons personally known to me; 300 for the general ones. Of course, they are re-circulated by some readers. Traditional publication of journals has been losing ground to online for the last ten years in the US.

I don't think history of mathematics is a paper in BA or MA in any university of India. If that be the case of no nursery for it, then is it only a research refuge for some mathematicians? The British universities have chair professorship in history of mathematics and Indian universities were initially modeled after them.

The Hindu thought is not historic in a sense -like, the Christians are; nor do they understand the conducive conditions for history of math! A journal or a research paper has to have a growing membership of the youthful students along with scholars. While looking at the complimentary issue, I really wondered if any high School or college student would find anything of interest and intelligible.

At his stage of life, you re-evaluate the mission of the GB and leave a new legacy. Of course, my views on India are bases on observations from a distance, while you are in the trenches. Thanks.

BSY: You are not thinking on the right lines. Ganita Bharati is the official bulletin of ISHM and it has its aims and objectives clearly defined. It is by and large meant for the specialists and consequently is bound to have a limited reading. I have tried to enlarge its readership within these specialists with effect from Vol. 28 as you might discover. Our purpose, and for that sake, of any journal worth name, cannot be just readership. You have undoubtedly a large readership of your reflections, but I am so sure, Ganita Bharati is not meant for most of them, nor do we aim to serve them. At the same time, your reflections in the present format, most of which I myself have enjoyed, cannot find a room in GB. Sorry, if I have not been able to put on board what I essentially mean. Regards.

I wrote: Don't feel sorry, particularly after 70! I enjoy writing these Reflections as Grattan-Guinness does it. A person like you finding some joy in reading my reflection means that I should continue doing it. However, they are never sent to you for publication purposes in the GB. I am really above this publish and perish rat race. At the same time, if by chance, any one of my reflections fits into GB, feel free to publish, or circulate. It is just like I enjoyed reading your ***Riemann Hypothesis*** article again, though it has been away from my present working.

A SAGA OF ELEMENTARY CALCULUS

Last month, the College dean shared his concern over high percentages of dropouts, withdrawals and low grades (less than C) in the development courses - identified one each from Biology, Chemistry, Math (181) and Physics. It is the first time that it caught the attention of a dean, though an old phenomenon. To my amusement, Geosciences, the fifth department in the College, does not have any developmental course. It is not surprising as Physics, Chemistry and Biology alone are traditional disciplines in natural sciences. Mathematics is their language; mother and nurse too!

Since UNLV is celebrating its 50th anniversary, it is befitting to give some history of **Elementary Calculus I (MATH 181)**. As trivia, its old course number, 121 was changed to 181 in 1990, and the prefix MAT to MATH in 2005. I first taught this course in 1977, though joined UNLV in 1974. Since then, it has been taught 11 times in a lecture format without requiring any math software. The following data is only drawn from my sections, but staggered over 30 years. They do suggest some trends. Besides, it is good resource for the new faculty members to align their expectations with UNLV students.

	Drop %	Full Time %	Sci/ Engg %	Math %	Freshmen %
1977 (Fa)	24	30	73	1	uk (unknown)
1984 (Fa)	22	29	50	0	36
1991 (Fa)	30	30	50	4	65
2002 (Sp)	38	uk	uk	uk	uk
2007 (Sum)	20	20	30	0	12.5

The following questions are pertinent to this data:

Is MATH 181 a freshmen course?

MATH 181 is not a freshmen course at UNLV! A sprinkling of freshmen students does take it, but by the time most of them finish it with a grade of C or higher, they are juniors! It is compounded by central advising that the basic science and engineering courses can be taken without calculus

49

first. The Catch-22 is that finishing calculus delays taking the major courses.

Where are the Math majors?

The students (7-8) who commit to math either they do by an elimination process or suddenly turned on by a math course or an instructor. Rarely, freshmen declare math as their major.

Who are the current stakeholders of Math?

They are no longer historical partners – Physics and Engineering. But, it is the growing number of students from pre-professional areas (namely, Biology and Chemistry) and business.

Drop Rates

No matter how one cuts off the grade roster, the percentages have not fluctuated much!

Work vs. Courses

UNLV students place equal, if not higher, priority on their jobs over the courses. Over the years, the number of MATH 181 repeaters has increased to nearly 50%!

However, it is better to work with this Las Vegas education culture than trying to change it. After all, UNLV is situated near the heart of world famous, Las Vegas Strip!

Aug 26, 2007/ July, 2014

COMMENTS

Same story (in general) nationally!! **Raja**

Satish, Hope you are fine. Our new semester has just started. Could you please remove me from your mailing list for the daily reflections? Thanks, **Mohan (Shrikhande)**

I wrote: Hey Mohan *Pyare* (means, dear), what is this?!

Your name is no longer on my mailing lists of *General Reflections* and *Hinduism Reflections*.

However, before I remove it from the list of *Mathematical Reflections* (average once a week!), I would appreciate your favor of giving me **two** good reasons for it.

I can indulge in asking for the reasons, since we are acquainted for over 40 years; though not as regular friends, but never on unfriendly terms either. A reason being that I am thinking of having them published very soon. If there is something that professionals like you find it unworthy, then I sure would thankfully take into consideration.

I leave for India next Saturday; hope your Las Vegas visit was refreshing. Thanks.

PS July 03, 2014. I never heard from Mohan again. I strongly believe in sharing my high moments of life with my genuine friends and relatives. Otherwise, what is the point of working hard in life? It is ridiculous to think that anonymous publications in journals are better than sharing them with people and professional near you. Likewise, I welcome nuggets of life from my matured friends and relatives, particularly when we are living in our 70s.

DISCOVERING MATHEMATICAL HISTORY
(A Reflective Note to Math Students in the University of Nizwa)

Introductory Remarks

A human life is defined by new discoveries and inventions, in every aspect - be that intellectual, physical, or spiritual. As a matter of fact, the common trait is the attitude to do something, never done before. It applies for success in academics, business, and politics. Therefore, you are urged to look deep into your lives, and find out where you stand against this measure. You alone are your best judge.

However, this opportunity will go a long way when this course is over - bachelor's diploma is almost in your hand, and you have become a matured adult in a society. Don't forget, that, as young you are, Time is in your favor, if you resolve to be a trailblazer

What is the research topic?

Together, we examine the contributions of the Omani men and women in the world of mathematical researches. Remember, I have come all the way from the USA, not just to teach you, but also learn from you - ancient Omani culture, and history. One can learn from any person, experience, and situation, if the mind remains open.

Since the geographical boundaries often change after every 50 years, and intellectuals, by choice or demand, always move to greener pastures, the neighboring countries Yemen and United Arab Emirates (UAE) may be included for the purpose of investigations.

Is this research paper mandatory?

NO; it is optional, but it is strongly recommended that every one gets its taste. It may turn out to be a transforming experience. The project will be due by the end of the 13th week.

Evaluation and credits

A research project/paper is evaluated on the basis of originality and the quantum of work. Hence one may earn anywhere between 3-9 % of the grade, at the top. If an individual paper is good, then it may be submitted for publication. In case, there are bits and piece of originality in a few papers, then they may be put together as a research paper for presentation at a conference, or sent for publication. In either case, your work will always be credited. Your name on a research paper may double the value of your diploma.

Why this topic on history of mathematics?

First, this is within the reach of everyone. Hard core research in mathematics, with your background, is too much to do by everyone in one semester. Besides, I have a stake in history of mathematics. Lately, I have been teaching courses, writing and publishing in History of Mathematics.

What is behind this research project?

Since all of you are seniors, ready to graduate within a year, and exposed to all basic mathematics topics, like algebra - classical and modern; analytic geometry, calculus, statistics, foundation topics, etc., this research project will mathematically connect you with Oman, and its immediate neighbors Yemen and the UAE.

How to do research?

First ask simple questions like: Names of the Omanis who have done PhD in mathematics? Are there women with PhD? What are their research areas? Just think, if there is no Omani PhD, then it may inspire you to be the first, or first in a certain branch of mathematics. You will learn about various branches and sub-branches of mathematics.

Where to look up for research answers?

Books: Yes; encyclopedias, reference books, history books may be a good beginning, particularly the ones in Arabic.

Journals and magazines: They are in the libraries of universities, college and public.

Internet: This research tool was not available 20 years ago. Use search engines like Google, Yahoo, and Wikipedia. Do speak with a reference librarian for getting information on **genuine research websites** for different subjects.

People: Don't underestimate your parents, high school teachers, professors in science and mathematics. Make a list of questions before meeting them, keep a log, and record the data afterwards. That is called oral history. You will get a first exposure to oral history.

Museums and Monuments: Since the modern books are only 500 years old, you shall have to be creative in finding mathematics from other sources like ancient monuments.

Read carefully my handout on *New Materials on History of Mathematics*, as distributed in the class last week. Make sure you fully understand it.

How to get started?

First decide, if you want to work in a team of two, or alone. A team of three or more may bring down individual efforts; so **a team of three or more is ruled out**. Through the end of the fourth week, just keep digging the sources listed above in the context of Omani mathematics. One or two topics will eventually emerge. Once you feel good about them, it means you have found a gold vein in a rocky mine. Come and discuss it with me, and I would clarify, help and **approve your individual or team topic**.

Updating

After all the topics are approved by the **end of the 5th week**, tentatively, every week, a 2-minute oral presentation(s) on updates will keep the collective spirit of research high. The oral presentations of work-in-progress will be a good exercise in communication ideas. English being new to you, it is essential to build fluency in its speaking and writing through every course you take. As you re-write your paper, your command over English will significantly improve.

Concluding Remarks

The benefits of this research paper are mathematical, personal, English (verbal and written). Since most of you want to become math teachers, this project would help, when you go out and teach. By the way, you may be the first students to do any mathematical research!

Feb 05, 2009 (Oman)/Jan 2010 [During Spring-2009, I was a visiting professor at a newly opened University of Nizwa in the Sultanate of Oman. Undergraduate research being unheard is this part of the world, I tarried to encourage the students as well as multinational faculty in my ongoing researches in history of mathematics. The following handout was distributed to the students in all three upper division courses that I taught there.]

EXTRA CREDITS ON MATHEMATICAL HISTORY

About two months ago, I circulated a similar handout on history of mathematics projects limited to the region comprising, Oman, UAE and Yemen. A couple of you have tasted what it takes to dig into such information. It is not simply going to internet and Google/Yahoo a few terms, make prints, and turn them in. That material may not be genuine, as so much trash is also is being thrown into the cyber space.

Again, reminding you that do not spend any time on it, if you are struggling with any course material which is of foremost importance. This project is generally for those students who love mathematics and want to go beyond A or B grades. Here are a few more ideas on history of mathematics projects.

1. Number 786. Is there anything special about it in Islam, or not? How and where this number comes from? Why some people believe or do not believe in its auspicious value?
2. University Nizwa was started in 2004 - its history is very brief. See, if you can dig some facts, like the following:
    (i) The names of mathematics **instructors** who have taught here since 2004; their nationalities, terminal degrees earned, institutions and years of graduation.
    (ii) All **mathematics** courses offered - excluding independent studies; in the order of semester.
    (iii) The **names of all the students** who have finished or are about to finish their bachelor's with major in mathematics.

Some sources of the data are: The offices of Mathematics Department, Dean of the College of Arts and Sciences, Human Resources, Registrar, Student Affairs and Vice Chancellor of Academic Affairs.

Gathering data from people require inter-personnel skills of communication too. So, you shall learn to be polite, patient and persistent. Incidentally, these are the qualities required for the understanding of mathematics!

Finally, the due date is the end of the 14th week of the semester. A team of **at most two** on a project is OK.

[Out of nearly 30 students only 6 - 7 students turned in the projects. From the perspective of history of mathematics, their value may not be significant, but this little exercise will go a long way in opening their minds closed for inquiries]

March 28, 2009/May, 2014

NIZWA WITHOUT HISTORY!

Generally, math faculty is not excited about teaching and researching in History of Mathematics (HoM). However, if the number of papers presented during the MAA, AMS, and ICM meetings is any measure of its popularity, then HoM beats the rest. Last week, in a curriculum committee of three, I was voted out by 2 to 1 for requiring HoM for BS at University of Nizwa (UN), located in an historic city of Oman.

It is essential to match a curriculum with the type of students and communities are around a new university. Hiring of the faculty must also be done accordingly. In Oman, the very first university was opened in 1986. There were hardly any roads and schools before 1970. The 90% of the UN students are girls who have no consideration for daily homework and regular attendance. Their priorities, being in a Muslim society, seem different.

Here is a slice of mathematical development of graduating seniors in the 12th week. My teaching assignment is of three courses. In a *Group Theory* course, not even a single student could correctly apply onto-ness of a function; recall any five of the ten properties of a vector space in *Linear Algebra*; or state the Fundamental Theorem of Arithmetic in *Number Theory*. On finding the current requirements of 74 math credits (excluding Calculus I) out of a total of 135 (academic torture), I initiated a few curricular changes.

HoM is not just about memorizing dates, names of theorem provers, or countries of their origin etc. It is like writing an essay on, say, lion. A six-year old child focuses on lion's physicals traits, a poet on its metaphorical qualities, and a zoologist on its anatomy, and so on. Calculus, and precalculus may be taught isomorphically in every land, but HoM is different in each land and culture. It is very pertinent in the Middle East with a past.

HoM comes alive when the students are engaged in mathematical past, present and future. HoM is mainly a movement of ideas. For instance, Euler (1707-83), born as Swiss, honored by the Paris Academy of

Sciences, but went back and forth to the academies in Germany under Fredrick the Great and Russia under Peter the Great - both patrons of science and math. It ties up with a broad question of necessary and sufficient conditions for the development of mathematics in a given political and religious system.

In the US, HoM is required for students working towards degrees in mathematics teaching in schools and colleges. But at the UN, it should be required for all math majors and minors. Above all, it is our professional obligation to teach history of our discipline. If we won't teach it, then who else would care, or can do a better job at it.

No one sets out a goal in life to prove a theorem in mathematics, but everyone wants to be remembered. We all have potential to make history, if it is cultivated in right doses throughout schools and colleges. Requiring history of mathematics course is also a major step in kindling indigenous interest in sciences and mathematics.

April 24, 2009 (Oman)/Jan, 2010

DARK SPOTS ON THE MOONS

An article, *'Write about Mathematicians in Non-Major Courses'* has appeared in the Nov. 2008 issue of the *FOCUS*, the news magazine of the *Mathematical Association of America.* However, it just came to my attention, as I have been catching up on things since returning to UNLV after a semester at the University of Nizwa, Oman. Its author, Karl-Dieter Crisman concludes that writing reports on the lives of top mathematicians motivates the students to study and appreciate mathematics. Lately, I have been integrating creative writing and history in all mathematics courses. The students can earn extra credits too.

This article struck a different approach in my mind. The students are to be given a list of mathematicians from several categories. A student may research into a life of his/her choice. Mathematical achievements won't escape the students. **However, I want them to pry, without any restrictions, into the non-mathematical recesses of their lives.**

What is the motive of this exercise? A human life is full of contradictions; no matter how great have been the achievements in one field or two. For instance, Michael Jackson, in death, is splashed all over the US and international media. From laymen to the presidents, kings and queens, all have been influenced by his music. Yet, there is a 'dark' side of his life when pedophilia charges were once leveled against him. Nevertheless, after a few decades, newer generations will only recall Michael Jackson as a musical genius.

My objective, being similar, is to find out dark, weak, or soft spots in the lives of mathematicians - from ancient Greeks, like, Pythagoras to the likes of Andrew Wiles, in the present. **What purpose will it achieve?** Mathematically inclined students are likely to be inspired by the obstacles overcome. Also, a majority of students, discouraged by mathematics, but on finding holes in the lives of mathematicians, may not give up on a particular course, or even in life in general.

My hypothesis is that every life is zero sum game on its vast stage of actions. A society or nation is selfish, as it idolizes the great minds that meet its ends. Unless we are able to identify with a life, however great in any sphere, at an approachable level, we are not going to take it seriously enough. The thrust is on discovering - how it airs out greatness hidden in others. Greatness never comes by following the footsteps of any over-achievers, for too long.

Conciseness in thoughts and expression are strictly stressed in write-ups. My norm of a length of a paper is 500 words, +/- 5 words. It is never mandatory, but the students are encouraged to give it a shot. Reading of the papers is joy, particularly, when a new nugget is brought out of a life familiar for years. On balance, a legendary life has to be critically examined for its foibles and frailties along with highlights.

Again, this writing exercise is a small step towards actualization of one's potential in early stages. We are all unique - it is a cliché. Life is not productively lived by simply knowing it. Strengths are discovered by overcoming challenges and hiking off the trails.

July 05, 2009/July, 2014

TEACHING BY THE RE-SEARCHERS

"Gottingen in 1846 was not the Mecca for mathematicians one would have expected it to be with the great Gauss in the chair of mathematics. Professors kept aloof from students and did not encourage original thinking or lecture on current research. Even Gauss himself taught only elementary courses." My thoughts immediately stopped in the track, as I just read it in a brief bio on Riemann, given in John Stillwell's book, *Mathematics and its History* (p 290; 2002). In fact, Riemann quit Gottingen and joined Berlin University, where he fully blossomed.

This year, I served on the Department Personnel Committee. Earlier, for three years, as Associate Dean, I was on the College Personnel Committee. UNLV's Math PhD Program is only four years old. While reviewing the applicants for promotion to the rank of (full) professor, it was clearly noted that 80-90 % of teaching was at 100-level courses; not even required of math majors! There is little teaching at upper division level, and a stray graduate course is taught in 4-5 years. In the US universities, teaching assignment is generally negotiated with the department chairperson.

Historically speaking, this 'benign' neglect of teaching by top researchers goes back to at least 200 years. Gauss, known as prince amongst mathematicians in the 19th century, was notorious for bad teaching. But UNLV is not Gottingen. One may argue that UNLV has the aspiration to become a Gottingen. But a lesson of history is that a Harvard or Stanford is created once in a century. Nevertheless, the public resources, specifically, must be spent judiciously all the time.

Since the PhD programs raise expectations in research, grants and scholarly activities, our college teaching load is commonly 2 - 2 – meaning, two courses per semester. In some cases, it is reduced to 1-2, and even to 1-1! Teaching is being looked punitive in today's academic culture. Math Dept had a viable MS program when I joined it in 1974. During the first year alone, I taught eight different courses (100-400 level) that new faculty members do not teach in their first five years.

Consequently, they develop anathema towards teaching and helping the students.

There is a disturbing conjugate side of research. The graduate/teaching assistants, doing MS/PhD, generally teach two 100-level courses (usually two sections of the same course). Over and above, they have to take 2-3 graduate courses towards the course work. Incidentally, some of them end up teaching more challenging courses than some 'research' faculty do! Well, I too, survived this academic 'slavery' for five years at IU (Indiana University). One may cynically argue that in a limiting case, exploitation sits at the foundation of every great civilization!

IU reminds me of several truly world-class mathematicians. Every semester, they taught one lower-division course and one graduate-level course in a cycle of 2-3 years - starting from first year graduate course to research seminars. Thus an instructor groomed and molded students into his/her area of expertise before accepting some for PhD work. Good researchers in math were good students first. **And, good students are only nurtured by dedicated teachers**.

Oct 29, 2009/July, 2014

COMMENTS

This is a wonderful testament to a balanced academic life consisting of teaching and research, and sharing our passion for education and training the next generation of leadership....it is certainly worth a departmental discussion. **Neal**

FIELD TRIP & HISTORY OF MATHEMATICS

When once-upon-a-time mathematician, gets interested in teaching history of mathematics courses, then teaching innovations are unbounded. It is not possible while teaching typical mathematics courses, which are bounded by traditional formats of instruction, contents, and students' expectations. After a gap of three years, I am teaching *History of Mathematics* (MAT 714) again.

Yesterday, I took all my ten graduate students to a campus museum having a rich collection of potteries, textiles and masks dating back to the Aztec, Mayan and Inca civilizations. The weekly assignment is to extract mathematics from the items on display that may be called 'DNAs' of lost civilizations.

Based upon my experience of numerous museums, the following suggestions were made:

1. Carry a small writing pad and a couple of colored pens.
2. Note down only one thought or idea on each page.
3. Jot down whatever connection strikes between displayed item(s) and math/science of the ancient era, when the item was supposedly produced (no issue of carbon dating).
4. Avoid rushing from one shelf to the other or from one display section to the other, as you explore hidden bridges. Absorption of knowledge is a slow process.
5. Take visual breaks and relax for a while. Let the delayed imagery of the items hit your consciousness in a secondary wave. It is a cycle of seeing; visualize connections, sit, and see.
6. For an hour or so, avoid discussing your observations with anyone. Let it become an entirely individualistic and internalized experience.
7. Watch your thinking going linear or 'Darwinian'- in the sense, that these ancient civilizations, having been conquered and annihilated, were generally primitive. The rise and fall of civilizations is like the birth and death of stars in the cosmos.

However, individuals, remembering the lessons of history, can slow down the decline of a nation.

8. The purpose of this museum field trip is to impress that history is not an arm chair discipline like mathematics. Here, personal opinions matter, and with varying emphasis.

9. A mathematics theorem, at times, may have more than one proof. But in history and archaeology, many distinct conclusions can be drawn from the same set of objects.

10. Try to extrapolate/interpolate the contemporary connections between math and similar objects with that of ancient times. This may open different windows of the mind.

COMMENTS

Sounds like a great idea and lesson--one I'm sure they won't forget!
Aaron Harris

THE FALL AND RISE OF TEACHING

During my first year at UNLV, I taught the following eight, **all different** courses (25 credits) in mathematics and computer science: **Fall 1974**: Math 096/Intermediate Algebra. (102 students), CSC 116/FORTRAN (16), Math 465/665/Numerical Analysis I (11), Math 473/673 Differential Equations I (10). **Spring 1975**: Math124/College Algebra (49), CSC 115/ BASIC (31), Math 459/659/Elem. Complex. Analysis (09), Math 466/666/ Numerical Analysis II (07). Thus, of the **seventeen** courses that were taught in the first five semesters, **twelve** were the new ones. Historically, till 1984, computer science courses were offered in Math Dept. These courses represent diversity in remedial, lower division, upper division/ graduate levels, and class sizes varying from 7 to 102.

For a number of reasons, I thought of 'bragging' about it and sharing this piece of both personal and departmental history. It also stems from a course in *History of Mathematics* (MAT 714) that I am currently teaching. For hands-on experience on research in history, the students are engaged in fine-tuning a history project of the Department. It was started in Spring 2007 with a different group of MAT 714 students for pitching in the commemoration activities of the 50th anniversary of UNLV. Math Dept may be the first department to have its entire history digitized.

There are other reasons too. Yesterday, during an open faculty meeting, Rosemary Renaut, a candidate for the deanship of the College of Sciences, voiced for tight evaluation for promotion and tenure at the department levels. She narrated the story of two cases under her watch as Dept Chair at Arizona State University. Coincidentally, as a member of the Department Personnel Committee, I am in the thick of midterm reviews of five colleagues, who joined the Dept in 2007. Any personnel evaluation is tough, but tenure evaluation is the toughest, as it impacts on collegiality for years to come.

The university academic culture has drastically changed since 1974. Presently, the teaching load of a new faculty member is 1-1 (means one course/semester) in the first year. It increments to 2-1 in the second year, and 'standard' 2-2 in the third year. In five semesters, each one of them has taught **seven** courses including 5-6 different ones. **Reduced teaching lowers the prestige of Teaching, but enhances that of Research**. The Service component of professional obligations, carrying 20% of performance weightage, goes way down in priority during the first few years.

Apart from reduced teaching, each new faculty member receives start–up funds to the order of $25,000 for research support, travels and computer. As a matter of fact, in sciences, the average start-up amount for setting up a lab is $400,000. Math comes for peanuts. The irony is that if a science faculty member is not tenured, then the lab equipments are all junked! In some cases, summer stipends are also provided. These research perks did not exist in 1970s. Research expectations were not high then, but they were not zero. I had a paper published and two accepted before getting tenure in 1977. I did not burn myself out for research – the papers just came out of my PhD dissertation. However, there was absolutely no pressure for grant writing.

In a free society, it is a simple fact of human behavior that pressure on research dilutes quality and integrity of research. It tend to inflate grades and encourages yellow teaching, grants get smaller and scarcer. Of course, the young faculty members may have to sacrifice their family life - thus creating long term societal problems.

Everything moves in a unique cycle. UNLV's academic stature has undergone significant changes - from Carnegie Master II in the 1970s to nearly Carnegie Research I in 2010. During this period of fiscal crisis and increased public accountability, teaching load may start rising. There is already 10-15 years of data on reduced faculty teaching loads, research productivity and funded grants. To the best of my knowledge, no national or institutional report has analyzed it for different Carnegie categories. In the meanwhile, a generation of faculty has passed out frowning upon teaching. A professional paradox is that good students are never inspired by instructors whose hearts are not into teaching. And, without a nursery of excellent undergraduates, a good crop of top researchers can never be expected.

Teaching load and research remind me of PR Halmos (1916-2006), a great mathematician, educator, textbook author and exponent of mathematics. *"We were kept busy, and we loved it. My teaching was 18 hours a week for several of the war semesters; the time I taught the "Super" ASTP course it was 21. Nevertheless, although my official affiliation with Syracuse lasted only three years, eight of my published papers list Syracuse as my address. Back then the days must have had 36 hours."*

This quote is taken from page 112 of Halmos' *Automathography*; *I want to be a mathematician* (1985). Due to economic depression of 1930s and WW II, the job market was in shambles. It took Halmos four years, after PhD (Illinois), to get his first regular job. He joined Syracuse University as an assistant professor in 1943. It did not have any graduate program then. Today, after 60+ years, the US has been engaged in two wars - Afghanistan and Iraq. Many major sectors of economy have gone upside down – another cycle.

April 15, 2010

COMMENTS

Hi Dr. Bhatnagar, I just wanted to reply to this particular reflection since I really take it to heart. I do try to read all the reflections that you e-mail to us students, but I have to admit that I actually made it a point to read this one several times. I especially love the following sentences: "A professional paradox is that good students are never inspired by instructors whose hearts are not in teaching. And, without a nursery of excellent students, a good crop of researchers can never be expected."

As I read this reflection, I had to stop at the end of that first sentence and read it out loud to fully enjoy it. I have wanted to be a teacher for as long as I can remember and I think the highest compliment I ever receive from students is when they tell me they can see that I really enjoy teaching. Sometimes I feel a little overwhelmed by my teaching duties, but from now on I'm going to remember what you've written in this reflection. I will always try to remember that my heart is in teaching and that in order to inspire my students, I need to try not to hide it. So thank you. - **Megan Austin-**

Satish, I had a few busy days at UNLV and was very very impressed with the interest of the science faculty in their potential Dean, I think I met about (at least perhaps) 50% of the entire science faculty. What a positive impression that provides. It is clear that the science college has faculty who care and want to work forward, as you note, to research I status. I see no reason why that will not happen.

I strongly believe that everyone in a department has a role to play, whatever the current status of that individual - research active or not. Tenure decisions have to reflect serious thought and consideration, but also very dedicated attention during the tenure-track years. I am sure you are dealing with these very questions at this time.

I had hoped to write and thank many for my visit, but now you beat me to it with an email to me. Thanks again to all of you, and thanks for taking me up on the suggestion to write to me.

Rosie/ Rosemary Renaut/Dept of Math and Stat, ASU/2010 -candidate for CoS Deanship

Thanks for sharing that Satish. As you know, I came to UNLV in 1984 when it was a different place than it is now, sometimes for the better and sometimes not. **Michael W. Bowers**, Ph.D. Executive Vice President and Provost

PERSONAL REMARKS

'SOME' HISTORY OF MATHEMATICS

Ten days ago, there was a half-day orientation of math graduate students in order to apprise them of departmental policies, faculty and programs etc. The faculty were encouraged to attend it and get acquainted with about 60 graduate students - admitted both in the MS and PhD programs. During introductory remarks, it is customary for the professors to talk about their mathematics expertise. I recounted that when I joined UNLV in 1974, the MS program was then six years old and today, the PhD program is six years old, as it was implemented in 2005. Since then, five students have finished their PhDs in different areas of mathematics.

Referring to my expertise, I added that my PhD was in Partial Differential Equations, but the current area of interest is History of Mathematics. Thus, I gave them a taste of Dept History101. Incidentally, history of UNLV's Mathematics Department was compiled over two semesters, as a collective hands-on history project during my offerings of *History of Mathematics* (MAT 714) in 2007 and 2010. Sixteen students had worked on various aspects of the Department – including students, faculty, administration, and curriculum.

Last Monday, during the first class of MAT 714, it was natural to expand on these remarks. A course on history of mathematics does not belong to any history department, as in such a department, mathematics knowledge of students and faculty is hardly beyond precalculus level. Also, if math professors won't teach history of their own discipline, then who else would care for it, and be qualified to do it?

One has to have a solid knowledge of mathematics for teaching a course on history of mathematics. At the same time, it does not mean that anyone knowing lot of mathematics is necessarily competent to teach a course on history of mathematics - far from it! One has to have a sense of general history, a macro outlook on life, and wide perspective on all areas of mathematics – belonging to possibly a rare breed of math generalists. On the contrary, since the 1950s, super specializations have been dominating every branch of science and mathematics in the US.

After five decades in collegiate mathematics, I have come to a conclusion that mathematicians really believe in the purity of their discipline. This is manifested in the ongoing debate between pure math and applied math - despite institutional pushes for interdisciplinary researches going on for the last ten years. In the class, I went on a limb in comparing purity in math with purity in Islamic rules. The deeper meaning behind the names - Taliban, Pakistan, Caliphate, Imamate, and Sultanate - is the same – religiously purest governments. Historically, each Islamic political system claims to rule under the purest laws of Islam, called the *Shariah*.

Historically, physics and mathematics were wedded and welded together for nearly 500 years. In my opinion, Newton is the greatest mathematician and the greatest physicist of all time. But the cracks in the edifice started appearing by the end of the 19th century. After the WW II, mathematical physicists started leaving the hallways of mathematics departments. Experts in continuum mechanics followed them soon after.

For nearly 50 years, mathematicians did not accept statisticians in their folds. According to a popular yarn, a mathematician told a statistician, "You guys don't have even a single theorem in your discipline. First prove a theorem, and then we shall consider you for admission into math fraternity." Well, the *Central Limit Theorem*, the first theorem did not help the statisticians much! Since the 1960s, either statisticians started voicing for their own departments or accepted to live as second-class citizens with mathematicians.

This story is repeated with computer science too. All mathematicians involved in computer science eventually left mathematics department. It happened at UNLV in 1984. In 1972, when I was at Indiana University, Bloomington, an established numerical analyst was forced to leave it. Math Dept was then dominated by the so-called, pure mathematicians - like Springer, Halmos, Brown, Azumaya and Stampfli - to name but a few. My thesis supervisor, Robert P. Gilbert, applied mathematician, who was appointed as full professor at the age of 35, eventually left IU in 1975 under similar circumstances. He joined University of Delaware as Unidel chair professor of mathematics.

I think a discipline worst treated off has been mathematics education. Mathematicians continue to look down upon PhDs in mathematics education as somewhat of 'lesser beings'. They may have reconciled to live with statisticians, but not with math educationists. On the contrary, the colleges and departments of education embrace mathematics education. They love them for their domain, 'education' in mathematics education - whereas, mathematicians do not love 'mathematics' in it! I strongly believe that a graduate program in mathematics education belongs to mathematics department, if graduate math courses are heavier in it. Needless to say, mathematics education provides the best funding opportunities for faculty research and mathematical services to the school teachers and community at large.

My excursion into mathematics education has been evolutionary in nature. After PhD, I was not passionate about continuing my thesis type research. As a matter of fact, in 1973 and 1974, I did not send my job applications to any PhD granting institution even though it was the worst employment period - caused by the first Middle East oil embargo. During the 1980s, I designed and taught several summer courses for the teachers. It was a very satisfying experience.

Interest in history of mathematics is a corollary of my growing passion for history, in general. Ironically, I hated history in high school, as taught then. Gradually, it started possessing my mind from different angles. To make it short, my forthcoming book, ***Vectors in History***, should technically qualify me to teach history of mathematics. If a recent formula of 10,000 hours per intellectual domain is applied, then I find myself standing on solid footing of history.

Towards the end the lecture, I threw an open question to the students on finding the US universities where graduate programs in history of mathematics are offered. I also added that in the Mathematical Association of America, the largest special interest group is of History of Mathematics – though no one may have earned a PhD in History of Mathematics!

Sep 05, 2011

COMMENTS

I agree that math education at times is treated quite poorly from mathematicians' perspective (which I don't understand as we are such advocates for the field in general). I also agree that a math education PhD would be better off in most math depts. with supplemental courses from education and Ed psych depts.

My undergrad degree from Utah State is a mathematics degree with a license for secondary education--from a math dept (not an education degree with emphasis in mathematics). I think secondary degrees for teaching should be housed in their respective content area department. Cheers, **Aaron**

WRITINGS ON HISTORY OF MATHEMATICS

Generally, a challenge in life is encountered from without, but on a fewer occasions, it is created from within. Whereas, an external challenge may be met to a degree of satisfaction, the internal one continues to simmer with dis-satisfaction for a protracted period of time. This observation is made in the context of a course on history of mathematics that I have taught on several occasions over the last 30 years. One of the challenges has been of finding a right textbook that meets my broad views on history of mathematics, as they are getting crystallized.

In the past, I have used the following textbooks in my courses: Howard Eaves (1911-2004) is an American doyen of history of mathematics. His two-volume *Great Moments in Mathematics* is a classic on mathematics highlights, but they leave a lot to be desired in between any two Moments. George V. Joseph's *The Crest of the Peacock* is the first book that focused on the non-European roots of mathematics. It is quite original in approach, but my head and heart yearn for far more than its pointed emphasis. John Stillman's *History of Mathematics* has excellent biographical sections at the end of each chapter, but the material gets mathematically heavier due to author's penchant for a unified approach towards a few mathematics strands. The most recent one is Roger Cooke's *The History of Mathematics* - used twice. It is the best of the lot, but like all other authors, Cooke also ignores a fact that any history of mathematics is a subset of history, in general.

After all, how can anyone talk of history of any one entity in total isolation of the world around it? One reason that I understand is that by and large, mathematicians are themselves not able to develop a holistic view of the world around them. Highly specialized researches in mathematics condition the minds towards tunnel and localized visions of life.

At times, I tend to think, like, Warren Buffet, that if I were raised in the US, I may have started compiling my notes on history of mathematics and turned them into a book long ago. To some extent, publishing is like public speaking. I have been trying to get over both of them. By joining a toastmasters club thirteen years ago, a lot of fear about public speaking has been overcome. Likewise, some *Mathematical Reflections*, started eight years ago, have been compiled into my first book published last year, at age 70.

The first book is a precursor to my book on history of mathematics where I boldly tread upon the impact of economics, political systems and organized religions in the development of mathematics and mathematicians. To the best of my knowledge, history of mathematics has never been approached from these angles, in particular.

From time to time, I like to change the textbooks of my courses for getting new perspectives on topics. It is no different from the movie directors seeking out younger female faces in the film industry. Incidentally, it rarely applies to the male actors. Last summer, while searching for a new textbook, I stumbled upon the *Math Book* by Clifford Pickover. It so piqued my interest that I adopted it as a lead book for a graduate course, MAT 714 (*History of Mathematics*). A glaring highlight of the book is that it is written by a person with neither PhD in mathematics nor history! Pickover's PhD is in microbiology. He has written more than 44 books, albeit to say, I am not envious of a kind of US 'publication mill'.

Pickover's scholarship may be fitting into a domain of 10,000 hours of expertise in a field. It is based on Malcom Gladwell's *Outliers*. Making history is nearly out of the domain of tenured conscious academic professoriate. In the US media - print, radio, television, websites and blogs, there are a great number of self-taught analysts and commentators who display intellectual prowess and boldness in interpreting social and political events. Some of them don't have even college degrees! However, they have invested years of their lives in self-studies and independent researches. In the US, there is always an alternate path to success. One can simply work one's way up- from the bottom to the top of a field, by sheer hard work, and without any college degree.

Also, the *Math Book* fits into the bit and bite world we live in at present - whether in reading, listening, cating, texting, tweeting, and so on. I was amazed to find its cost under $30, whereas, a typical math textbook is around $150! The *Math βook* has 250 glossy pictures on each page and 400-word write-ups on mathematical topics, on the opposite page.

Nevertheless, the *Math βook* suffers from the chorionic disease of wholesale turning of the pages of time - starting from 150 million years ago, to jumping forward to 30 million years ago, and to 1 million years ago, and so on! No apologies! Either, it is arrogance and ignorance, or extrapolated Darwinian approach. A spirit of 'intellectual mafia' seems to be in play. It perpetuates an approach which involves generalizing strictly local data decoded from an archaeological artifact. For instance, in Section 5, the most ridiculous conclusion was to associate some random markings on a piece of bone (*Ishango Bone*) to the 'menstrual cycle of a 'woman' - lived 18,000 years ago. Sheer nonsense! The obsession of the Americans with public expression of sexuality and crimes are reflected in such intellectual derivations – bordering deviations.

Nevertheless, students and I, together, will critically scrutinize this book throughout this semester. I am now certain that my compilation of *Reflections* on **History of Mathematics** will provide fresh insights into this monolithic subject. A reason for the dearth of innovative textbooks in history of mathematics is that such a course is not required for math majors in most US universities. It is only required in some teacher training programs. Math textbooks are written when there is a good captive market for it, or when one just wants to be recognized as author of a book.

Personally, I believe that a course in history of mathematics is essential for every math major – graduate or undergraduate, and potential school or college teacher. Also, it is a corollary of 4-6 credits of History requirement in the General Education Core for a bachelor's from UNLV.

Sep 11, 2011

AN APPETIZER IN HISTORY

A course in history without any hands-on experience is like studying science without doing lab experiments. For obtaining historical information, the researchers, who used to run to the libraries, now surf the internet. There is little digging of new facts by hand shovels; but there may be new interpretations of the old facts. Ever since my interest gravitated towards teaching courses in history of mathematics, I decided to bring some radical changes. In the last two offerings of *History of Mathematics* (MAT 714), an assignment was given out on the very first day: write a brief report on your any great grandparent - there are eight of them in all.

You can't go online and search a great grandparent of every person in the world. Also, the odds of a 90-year old person in full communicable senses are very small. So far, no one has known a great grandparent first hand. This realization alone shakes the minds out of its monotonous routines. Life appears to float on magic sheets that the moment you turn it over, it is all erased and is ready for new writings on it!

In many cultures, records have been passed on orally from generations to generations. The written records of ancient civilizations of Babylonia, Egypt, India and China, were commissioned by the rulers and they were transcribed on metal plates, stone steles, and compressed pulp of the trees and weeds. These writings, being beyond certain means of even the aristocrats, were elitist, but authentic, nevertheless. Once the mass printing took over, affluent classes and individuals started preserving their records for posterities, and for a piece of immortality. For the have-nots – whether individuals or nations, written records thus became sacrosanct. However, the oral and written records are equally reliable or unworthy in the present digital age. For years, I have wrestled with this thought.

About the reports, one student found that her great grandmother was a teacher in Nebraska, and she also found a document on math syllabus of the 8th grade. Its math problems would beat down some present college students. Another student's great grandfather went on a church mission to India in 1876. It touched me personally. Some students contacted their relatives in Poland, Turkey and Afghanistan and put out composite pictures of their ancestors. It revealed how organized religions keep people closer. Despite initial skepticism, everyone enjoyed it.

This exercise serves two main objectives. Finding **ancient** math or its mathematical evidence is far more difficult than tracing one's great grandparent, though everyone has one! The statement that mathematics and sciences flourish in organized and prosperous nations is irrefutable. Math has to be extracted like information on a person from its DNA.

Mathematical thinking is innate in humans. The Greeks, Chinese and Indians brought them out, as their ancient treatises point it out. But the world credits the Greeks only. The reasons are political. The Chinese and Hindus lost their heritage during their long periods of colonization. The first thing the colonizers do is to destroy the past of the colonized nations. History could be cruel or fascinating – depending which side you are on!

Sep 15, 2011

COMMENTS

Dear Satish, Thanks for this Reflection!

A meeting of the Ramanujan Mathematical Society is scheduled to take place at Allahabad during the first week of October 2011. The programme includes a symposium on History of Indian Mathematics. While I don't really have any expertise in this subject, all the same I am going to be one of the speakers. I have decided to speak on the History of Mathematics in Punjab. With your deep interest in the subject, as a part of the preparation for my talk, I will appreciate receiving your recollections/views on this topic. Best wishes, Sincerely, **Inder Bir**

DESSERTS IN A HISTORY COURSE

In life, either one over-estimates one's achievements, or under-estimates them. Rarely, one is right on the dot. It may be due to a variation of the *Heisenberg Principle of Indeterminacy*, or a paradox of an observer observing itself. The point is that any indifference to one's work, particularly, in the youthful years, is inimical to creativity in the future. I suffered from this psychological hang-up while growing up in Bathinda (India) of the 1950s.

This train of thought has been knocking my mind for the last two months while I was preparing to teach a graduate course, *History of Mathematics* (MAT 714). One of the evaluation criteria is that each student must work on an individual project of 'constructing' history. The project could be 'mathematically' local or global. It must be chosen and approved by the middle of the semester, and finished by its end. Above all, the student's heart must be into the project.

The idea of a project struck my mind during Spring-2007, when I first taught this course. UNLV was preparing for the 50th anniversary of its founding in 1957. I said to myself: why not to compile a 50-year history of my Mathematical Sciences Department for its posterities of students, faculty, alumni, or curious pubic. The Department was parsed and parceled out from ten different angles – like math majors, faculty, various courses, gender etc. There being ten students in the course, each one picked up the topic according to his/her taste. The end product collectively turned out pretty good.

The most interesting finding was the year of its founding. It is 1957, according to some university records, or is it 1956, according to Emeritus Mathematics Professor, Malcolm Graham (1956-1985), who was hired in 1956!? The feature of revising or correcting facts sets history apart from mathematics. History is revised and facts are changed solely according to the collective and political will of the people. In the world of mathematics, a wrong theorem never gets any traction. Once a theorem is proved; it means moving forward. No one is credited for proving a theorem again and again!

The students learn a lot about what it takes in the way of finding facts in history vs. proving a theorem in mathematics. For instance, while interviewing people, going to the archives of the Registrar's office, or to the Special Collections of UNLV Library, they face bureaucratic hurdles, various privacy laws, and, no less, realize the importance of communication skills.

After a year, I realized that another round of collective research was needed in a few spots for a finished product. In Spring-2010, those areas were reassigned to six students, who took this course. Now, Mathematical Sciences Department is, perhaps, the only department that has a reasonably comprehensive history. What a feeling of collective achievement!

Now, all the records will be shortly deposited with the Department Chair. They have been lying in my office for a while. It will require scanning and saving in a digital format so that it can be accessed and posted on a website, as and when it is needed.

The scholars are known to be armchair thinkers. In the digital era, they are topped by their laptops, iPads, and iThings. However, in any discipline, some research problems can only be solved with the sweat of both body and mind. That is a motivation of doing a project. This semester, four out of initial five students are taking this course. Two of them are full time teachers, and they are working on the histories of their schools with respect to math curricula, faculty etc. The remaining two will put together some mathematical histories of Afghanistan and Poland, as they have some resources and roots there.

The joy of teaching this course comes from the readings of weekly reports and final projects. I am sure, when all is said and done about this course, students will remember what they would actually discover with their efforts. We all forget what is done routinely. Above all, it is all about making and tasting history - though it may be at epsilon-delta level to begin with!

Oct 13/Nov 13, 2011

COMMENTS

Satish, I'm glad that you've seen the light and become a historian!! Seriously, this sounds very intriguing. **Michael W. Bowers** (Provost and Executive VP)

PERSONAL REMARKS

UNEXPECTED HISTORICAL NUGGETS

Call it a desire to innovate, break out a new trail, or try a new approach - that is how history of UNLV's Department of Mathematical Sciences has developed or evolved over a period of five years. It is essentially a byproduct of teaching a graduate course in *History of Mathematics* (MAT 714) twice. It is a required course for MS in Mathematics with Teaching Concentration - one of the four concentrations. As soon as it was finalized that I was to teach this course in Spring-2007, it was decided to include individual hands-on history projects.

My thinking comes from a belief that a science course without any lab experience is not complete for developing scientific thinking and instilling a sense of its appreciation. Likewise, history at the graduate level is not just interpreting historical facts, but also questioning their very genesis etc.

As the semester started, I regularly shared the idea of small history projects with my students. Generally, the students are conditioned to read history without questioning as to how some of the facts have come about in the first place. I convinced them of the long-term benefits of the projects, particularly when any piece of history would ever be studied. Also, I was convinced that the nature of projects should be such that the students do not get help from any online or typical library sources. Instead, the emphasis was on old-fashioned walking to the sources, talking with resourceful persons, and collecting bits and pieces of oral history.

With these broad boundary conditions, as I was working on history projects to be assigned to each individual student, I realized that year-2007 was to be the 50th anniversary of the founding of UNLV. Collectively, we decided to pitch our contribution into its history. Once that stuck my mind, it was easy to parcel out every aspect of the Department of Mathematical Sciences into small manageable research projects.

The class of 2007 worked on the projects dealing with the curriculum, students, faculty, and administration. I was quite satisfied with the entire product, but it required more work in some areas, and research into additional topics, which had been missed out in the first round. So, when I got to teach this course in Spring-2010, some major gaps were filled in by the new class.

Well, I take pride in the fact that no other UNLV department may have compiled such a history. All the projects are put together in two 3-ring folders - each two inches thick. They have remained in a corner of my filing cabinet for over 3 years – waiting, perhaps, for a new home. After all, what good is this collective effort, if the material is not made public? Two places that I think of are the Department library and Special Collections of UNLV's Lied Library. The students have used both of them. I am not a technologically savvy person; however, the material could be scanned or digitized, and placed there.

Listed below are the names of the students in alphabetical order, who have contributed their projects: Meagan Austin, Brian Carpenter, Nathan Dabalo, Maryanne Griffin, Marjan Hakimian, Aaron Harris, Tina Larios, Lee Karatich, Scott MacDonald, Matthew Michaelson, Kristine Paulsen, Matt Philips, Zachary Porter, Jeff Reeves, Whitney Rotert, Joel Schimeck, and Timothy Wiseman.

In order to give a brief idea of the contents, the following are the titles of some of the projects that were taken up: History of Calculus I at UNLV, UNLV Math Undergraduates 1957 - 2009, Demographics of UNLV Mathematics Majors and Graduates, Math Faculty and Chairs -including their educational, ethnic and geographical backgrounds; some part time faculty too is there; Female Faculty of the Department, Fifty Years of Undergraduate Math Majors, Graduate Students of UNLV Math Dept., Remedial Math courses, History of College Algebra, History of Trigonometry and Analytic Geometry, A History of Precalculus.

There is a whole lot more under each of the heading. Sometimes, more than one student has worked on a topic bringing richness and new angles of approach.

Aug 15, 2014

COMMENTS

This is a good example of how there can be interesting overlap among the academic disciplines.
Francis

PERSONAL REMARKS

SECTION II

HUMANISTIC SLICES

[The following is the text of a letter to the editor of the *College Mathematics Journal* (CMJ) published by the Mathematical Association of America. It is 25+ years old, but the interview with mathematician, Mary Ellen Rudin (1924-2013) was so well reported that I must have quoted it many times during teaching in different contexts – including women in mathematics, numeric understanding of children with special needs, profession vs family, nature of mathematics, and so on. It merits 5-star inclusion in the book]

ON INTERVIEW WITH MARY RUDIN

Editors are to be congratulated for changing the format of the CMJ by publishing an extensive interview with Mary Rudin in its volume, 19 (2). There are several reasons for liking it. Personally, it struck a different note altogether. The whole interview with Mary Ellen was a breath of fresh air amidst 'mathematical stuffiness' in general.

During 1987, I happened to read three reviews of Halmos' book, *I want to be a mathematician: An Automathography*. The intersection of all the reviews is that Halmos did not include anything in his book which had no mathematical bearing of some sort. As one reviewer aptly put it, that after reading the whole book, one wonders whether Halmos even had a wife and children.

On the other hand, Mary Rudin's mathematics appears to have thrived amidst her household chores, children and husband. Yes, her husband, Walter Rudin (1921- 2010) was a rare combination of a distinguished researcher and writer of high quality graduate textbooks. Since we all have role models and mentors in life, therefore, my comments are only directed towards attitudes surrounding mathematics. No qualitative or quantitative comparison between mathematical productivity is intended, or implied.

Moreover, this is the only item from any mathematics journal which my whole family has read it. My wife has studied philosophy; daughters,

English; and son, mathematics and computer science. The interview impressed me with a laudable fact that a wife can pursue her intellectual pursuits without being sucked in a rat race for a job, provided husband's income is adequate.

My daughter has a two-month old daughter, and feels motivated to go for PhD. Raising a family and deeper contemplation upon ideas - whether in mathematics or in literature, are not mutually exclusive. If there is no contradiction in being both a housewife and a mathematician, then there is absolutely no problem being a wife and any other professional.

July 04, 1988/May, 2014

LAS VEGAS & SAUNDERS MACLANE

Today's early morning e-mail from a neighbor prompted me to click on an internet link to an obituary appeared in the **Boston Globe**. Its heading read: **"Saunders MacLane, developed key Algebraic theory; at 95."** My instant reaction was, "My God, this man was doing fundamental research in mathematics at 95!" But this reality lasted only for a second, as the title was misleading. In his death on April 15, 2005, MacLane may have created a record as the longest living and active mathematician of the 20th century!

While replying to my Bostonian neighbor, I said, "MacLane is the one man responsible for not letting the annual Joint Mathematics Meetings of the AMS (American Mathematical Society), MAA (Mathematical Association of America), and other mathematics organizations take place in Las Vegas! Very stubborn man to the end."

I continued to muse over my comments. After all, how can you crack a long standing mathematics problem or develop a new concept without persistence and perseverance? Never! One's being stubborn in dealing with colleagues may not be collegial - a positive quality. Nevertheless, in certain areas of 'pure' mathematics, there may be a correlation between original research and stubbornness. It is not a litmus test, though in my small sample, it works with 95% accuracy!

During the first and only Joint Meetings held in Las Vegas in Jan 1972, MacLane pushed a resolution that has ruled Las Vegas out as a conference site! I was shocked to discover it, when my interest in the Meetings grew in the 1980s. A couple of math bigwigs in the organization essentially told me, "You can forget Las Vegas as long as MacLane is alive!" Once I wrote him a very persuasive letter on this issue. He graciously replied it in his long hand, but refused to change his mind.

One may talk of his legacy - both mathematical and non-mathematical. They are two separate things. One does not have to study **Category Theory** to know its impact on many branches of mathematics including computer science. His influence both on the US and international mathematics organizations is deep for the number of people who know him directly (his colleagues and PhD students at Level 1), and indirectly (his students' students, Level 2, 3, 4, and 5!). To give an idea of his hold over mathematics organizations, I never got a satisfactory reply from any officer of the AMS/MAA of my written suggestions for holding a Joint Meeting in Las Vegas! It was essentially a one man crusade that I loosened up five years ago.

Yes, it was five years ago, when I last saw MacLane at a Joint Meeting. He spoke at a panel discussion on *Philosophy of Mathematics*. People were standing wall to wall in a big hall to hear him! He had the lecture notes written up and rolled in yellow sheets. Most of the time, he rambled on. In fairness, that was a great mathematical performance at 90!

MacLane must have been asked about a secret of his longevity. I think, had he not re-married at near 80, he would have died of loneliness (A line from a work of George Bernard Shaw's essay)!

April 23, 2005/June, 2014

[PS: For its intersection with general reflections, its variation is included in the *Scattered Matherticles*: *Mathematical Reflections*, Volume I (2010). A part of MacLane's legacy can be measured by the fact that Las Vegas in nowhere on the list of future sites of the JMM. My interest in attending JMM has also diminished from annual to once in 2-3 years. Thus, my campaign for Las Vegas site has become passive. However, I find it ridiculous when the JMM are held in Boston and Baltimore in the cold month of January. For the visitors, Las Vegas weather and convention facilities are simply incredible. Whereas, Las Vegas may be the last frontier of new applications of mathematics!]

COMMENTS

I remember picking mathematics as my college major when I was 18 years old. I decided at that point that my goal in life was to be the best mathematician that I could be. Looking back, I see that I often let life get into the way of that goal. Five years ago, when I turned 50, I remember thinking, this is as good as I am ever going to get. I compared myself to some of the mathematicians that I know and admire, and got very depressed. But I couldn't stay depressed.

The fact is, however mediocre a mathematician I am, I still love mathematics. I still lose sleep at night working on problems. Some are my own invention and many more invented by others. I still carry a notebook around with me full of mathematical "doodlings" and I am always pulling it out and "doodling" into it some more whenever I find the free time. I know that I am hooked on the joy of discovery and I hope that, should I live to be 95, I will still have that desire to know and discover burning in me. **Steven Gregory** (My MS student in the 1970s)

Satish, Mike Golberg, Looy Simonoff, and I were on the local arrangements committee when the joint meeting was here in Las Vegas. We were not pleased with the motion to never return to Las Vegas in this (1900s) century. It was a good meeting. People liked it. I have had little regard for MacLane since. Being a decent human being is more important to me than being a great mathematician. Las Vegas has flourished and MacLane is dead. Who won? **Paul Aizley** (Emeritus Mathematics Professor and three-term Nevada Assemblyman)

Sad to know of his demise. MacLane's Mathematics, Form and Function were how I spent time outside classes during my final year of MSc... **Shankar**

Thanks for including me on the famous reflections email list. If he had not remarried at the age of 80, he might have lengthened his record beyond 95! **Sangitha**

A TALE OF TWO MATHEMATICS

During Dec 26-29, 2005, the 71st annual conference of the Indian Mathematical Society (IMS) was held in Roorkee, India. Last week (Jan 11-15, 2006), it was the Joint Mathematics Meetings; the 108th meeting of the American Mathematical Society (AMS), 85th of the Mathematical Association of America (MAA), and other small mathematics organizations, held in San Antonio, Texas. Coincidentally, I happened to attend both of them. So, it is natural to reflect on this double experience never had before.

I became a life member of the IMS during my two year sojourn (1980-82) in India. Except for the first year, never any communication has been received from the Society! The IMS has no permanent office like the AMS and MAA have them. Its virtual office moves physically with the election of new officers. Despite the fact that Indians today are world famous as IT experts, but the IMS maintains no website!

Like many learned societies in India, the IMS was founded in 1907 by the British scholars coming on various assignments to India. Gradually, Indians took it over after independence in 1947. It is interesting to note that the AMS, founded in 1888, has over 28,000 members and the MAA, in 1920, has 30,000. In 1976, I switched my membership from the AMS to MAA.

The social clubs and professional organizations are the hallmarks of the organized western mind. The Hindus turning into highly individualistic, by and large, their organizations flounder away eventually. There are varied annual membership dues for the AMS and MAA, but none for the IMS except its lifetime membership. It is sheer paradox, that whereas the number of students doing master's and PhDs in math, and of faculty teaching math in colleges and universities is perhaps four times that of the US, yet the IMS has hardly grown in 50 years. The President of the IMS told me that for years the attendance at the annual meetings has been around 100!

My first memory of an IMS meeting goes back to 1960 when I attended it in Chandigarh as a student volunteer – then doing MA (Part II). My second meeting was in 1986 while on sabbatical leave in India. It was held in Jaipur, and presided by my teacher and mentor, Professor SD Chopra. The Roorkee Conference was presided by Sarvajit Singh, known to me since 1965. Both of us were PhD students of Professor SD Chopra in Kurukshetra University (KU). However, I left KU in 1967 after two years without finishing the thesis.

It was heartening to watch Sarva Jit give the presidential remarks followed by his technical address. One feels honored in watching a friend being honored! The entire gathering was accommodated in the auditorium of the mathematics and physics departments of IIT Roorkee. As a historical footnote, the present IIT Roorkee was started in 1847 as the first engineering college in the British India. Yes, the entire conference deliberations, paper sessions, and stay of the delegates were on the campus.

In contrast, the San Antonio Meetings was held in the San Antonio Convention Hall. It was attended by over 5000 mathematicians, students, some spouses from all over the US and world. It is the greatest spectacle of mathematics on earth! There was a Press Room for daily releases. Thousands of papers were presented in hundreds of sessions in addition to various committee meetings, workshops, exhibits, and publishers. The high-tech Employment Centre alone covered an area of 300,000 sq. ft. It draws employers and job seekers from all over the world.

The mission of the Joint Meetings is to promote mathematics at every forum with teaching innovations, integrating technology, curricular reforms, publications, and latest research in every conceivable area of mathematics. Currently, my favorite topics are ***Sports and Mathematics***, ***Arts and Mathematics*** and ***History of Mathematics***. There were 4-5 booths in the Exhibits promoting mathematical concepts printed on the T-shirts, hewn in stones, chiseled in wood, and designed in jewelry. One can soak mathematics of whatever taste and style one likes. The eyes cannot see it all, and mind cannot comprehend it either!

The delegates, staying in about ten hotels, boost the local economy. They are welcomed by every establishment. Convention is a business and the Americans are best at holding it. Each mathematics organization has a fulltime, Executive Director and staff who run it with the policies laid out by its elected Board. Members pay dues and in return get the services and benefits of a professional organization.

In contrast, the IMS seem to be spinning around a couple of individuals who are prominent by special VIP badges on their lapels and a cluster of individuals hovering over them. It is a typical Indian style where an individual in office works to enhance his/her image and not build the organization. The western mind generally leaves the office stronger than before. It essentially defines the functioning styles of east and west.

One wonders as to how the expenses of the IMS conferences are borne when there is little source of regular revenue. Since the entire show involves about 100 people, the expenses are relatively small too. The host institution bears the bulk of it. Being vacation time on the campus, the classrooms, auditoriums, dorms, and guest houses for the VIPs are freely available for the conference activities. Also, some state and central government agencies pitch in healthy financial support.

The highlights of the IMS are the refreshments during tea/coffee breaks, luncheons and dinners – all included in the registration fee. It provides an intimate atmosphere for interaction. To top it all are free cultural programs in the evenings by the local artists. That involves a tremendous organizational work on the part of the host institution. In the US, the local institutions may have some advisory role, but the organization of the conference is mostly done professionally. Such a great event has a template that the core staff uses every year, thus improving and adjusting it according to local conditions.

I attend these conventions for a couple of reasons. Number one is social, in meeting old friends and making new ones. Besides, presenting a paper, or organizing a session, as I did this year, on ***Philosophy of Mathematics***, I love to attend various invited addressees for getting global views on mathematical issues, prize functions to applaud the new stars of mathematics, and technical papers to keep my mathematical muscles in shape. Anything else is bonus! At one time, I used to stay for all the four days, but now two days suffice.

It is a universal human urge to see and meet a legend in one's field. A great American mathematician, PR Halmos (I took a course from him) once remarked, "One must support a math organization by regular attendance." Halmos has not been attending it for the last couple of years for health reasons at 90. But I did spot 80-year old agile and silvery Peter Lax, the 2005 winner of the Abel Prize (equivalent to the Nobel Prize in money). In a huge networking area, I enjoy taking notes and observing people associated with mathematics.

The aspect of obligatory attendance of the conference was absent in the IMS. A couple of my former teachers and friends who had served as presidents of the IMS did not want to travel a distance of mere 100 miles to attend the conference. At the IMS meeting, either you are somebody as an invited speaker, or nobody. Can this culture change? It is an open question that Indian mathematicians should be able to tackle.

For many years after settling in US, I did not appreciate the power and benefits of professional organizations. If the US is leading in its world class math departments, PhD programs, innovative courses, and cutting edge research, then it is because of the people working together through the AMS, MAA, and other professional organizations.

Incidentally, PR Halmos and his wife donated nearly 3 million dollars to the MAA a couple of years ago. Voluntary donation to professional organizations beside regular dues further sets the west apart. The Indian mathematicians have a long way to go in giving donations to promote their own discipline.

The US mathematics convention is like a mini United Nation, as one can see people from every ethnic group and nationality. I saw relatively more Indians in the San Antonio convention than at the Roorkee conference! At the IMS meeting, only two Sikhs and one Muslim were there - no way to identify Christians.

On noticing a Muslim participant feeling alone, I introduced myself. He turned out to be from Iran, and came to India for doing PhD from a university in Pune! That speaks of high caliber of individual Indian professors and universities to attract graduate students from overseas. As a sidebar observation, there is a statistical correlation between the religious beliefs and the study of abstracts areas of mathematics.

Jan 16, 2006/June, 2014

[PS: Due to non-empty intersection of ideas, a version of this reflection has appeared in the *Scattered Matherticles*: *Mathematical Reflections, Volume I* (2010).]

COMMENTS

Satish, Paul Halmos was not a great mathematician. Rudin referred to him once as the greatest second rate mathematician. I agree with Rudin. Someday I will tell you some Halmos stories which will change your mind. Best, **Bob** (Robert P. Gilbert, Emeritus Unidel Professor, University of Delaware, Newark)

Question: Is IMS still run by Singal trio - M. K. Singal, his wife, and her sister Shashi Prabha Arya? **Subhash Saxena**

Dear Satish, Your incisive analysis of the comparison between the functioning of IMS and AMS was very interesting to read. With regards, Sincerely, **Inder Bir**

WHO IS A GREAT MATHEMATICIAN?

This question is an offshoot of my recent *mathematical reflection* in which I mentioned PR Halmos as a great American mathematician. A few comments were received, that in turn re-set my mind to contemplate over it. After all, I have already spent 45 years in the service of mathematics. Question: How to measure the greatness of a mathematician?

Last night, I witnessed Kobe Bryant scoring 81 points in a game. A commentator described him a basketball machine. The basketball fans will remember him a great shooter. In every sport, the players are complete in more than one aspect of a game. Nevertheless, in any team sport, an individual performance comes after the team performance.

That is not the general perception of a great mathematician. The modern history of mathematics is mostly Eurocentric. The great Hindu and Chinese mathematicians of the past are isolated in the sense that little is known of their contributions, and schools of thought. Their long colorizations have made a parking lot of their heritage and monuments. In Europe, until, the middle of the 19th century, most mathematicians were essentially mathematical physicists, or applied mathematician in the lexicon of 20th century mathematics.

Taking recent history as guide, the world remembers those who have enhanced the progress of mathematics by proving new theorems, or solving great problems. Ironically, the world knows nothing about some solid mathematicians who spent all their lives on very difficult and unsolved problems like Riemann Hypothesis or Goldbach Conjecture. Also, mathematicians, no matter how inspiring they have been as teachers, remain unrecognized. That is a global view of professionals in mathematics today.

I go a step further – beyond research and teaching to define a mathematician. Just like, there is a woman behind a successful man; there is a department chair behind a great mathematician or a group of researchers. It is the department chair who is able to pull various resources together in creating a needed nurturing and nourishing environment. For individual research, teaching and service, he gets support from the college dean in particular. The communication and administrative skill is rare amongst math professors.

Another thought was to compare mathematics to electrical power. Electrical engineers are involved in three stages; power generation, transmission, and distribution. By and large, mathematics has three components: generating new math by proving new theorems, transmitting math by teaching and cultivating new generations, distributing mathematical knowledge by textbook publications, or applying it in science and technology.

The question of the greatest mathematician is not like the highest peak of a mountain range. In the US today, men and women, schools and colleges, teams and institutes are professionally ranked every day. Mathematics departments are ranked too. Though there is no Nobel Prize in mathematics, but the Fields Medal, Abel Prize do provide a measure of mathematical research only. The annual teaching and service awards of the MAA and AMS have put national spotlight away from research. But they are limited to the US only. Teaching is so culturally dependent, that the comparisons are difficult across nations.

In the US alone, mathematics is taught in over 5000 colleges and universities. Who is a great mathematician of a year/decade in terms of research, teaching and service? It is a good question. Beyond its discussion over drinks, its answer will remain 'local'. It may be least incontrovertible, if the term, mathematician is qualified, as research mathematician, teaching mathematician (like student-athlete), or evaluated composite, based on 40% research, 40 % teaching and 20% service - the way math faculty are evaluated for tenure, promotion and merit at UNLV.

Finally, it is my time to declare an overall (40:40:20) best local mathematician that I have personally known and observed as a student or colleague over the years. They are Hans Raj Gupta (1902-1988) of Panjab University, Chandigarh, Som Dutt Chopra (1915-1987) of Kurukshetra University and George Springer (1924- present) of Indiana University, Melwane Anand (1962- present) at UNLV.

Jan 2006/June, 2014

[PS: This is significantly modified piece from a reflection that has appeared in the *Scattered Matherticles*: *Mathematical Reflections, Volume I* (2010)]

COMMENTS

All what you write now and even wrote earlier on the subject is, the least I should say, subjective. Think of 'E.T. Bell writing Men of Mathematics' again! And naturally expanding his contents. Halmos himself assesses him in his autobiography for readers to have an idea. However this is again subjective, and I do not agree with him. Regards, **B. S. Yadav**

MARBLES OF RESEARCH

When it comes to problem solving, **'Two heads are better than one'**, is a popular cliché in the US public life. It applies not just for figuring out mundane problems, but in fundamental researches too. Mathematics, in general, is no exception, though research in 'Pure Mathematics' is still regarded a highly individualistic intellectual activity.

While growing up in India, a belief set in my mind was that basic math research was a lone man's pursuit! It is indeed more of a team work. The US life has opened my eyes. Intellectual collaboration between two individuals is like free for all popular fights using legs, fists, and even certain objects. This is witnessed amongst the US researchers. They argue over a problem, but at the end of the week, they drink over it.

This cord struck my mind yesterday while reading a brief article on the Bernoullis who dominated the world of mathematics spanning 3-4 generations (1650-1800). Out of a total of eight Bernoulli mathematicians, three all-time greats are; two brothers Jakobi (1654-1705) and Johann (1667-1754), and Johann's son, Daniel (1700-1782). Jakobi, the first child and Johann, the 10th, were born in a wealthy business family of Switzerland.

The lives of the Bernoullis are filled with stories of competitiveness and jealousy between brothers, cousins, uncles and nephews, and ultimately between father and son. During the beginning years, the youngsters did learn mathematics from their elders, but as professionals they became fiercely independent and arrogant. It is like the Mughals ruling North India for nearly 200 years (1525-1725). The heir apparent brothers were killing and fighting all the time for the crown of Delhi. The Chinese revolutionary, Mao Tse-tung (1893-1976) said, "The power comes from the barrel of the gun." It applies in any walk of life, if one aspires to be at the Everest of its mountain.

Twenty years ago, a white American physics colleague at UNLV frankly confided in me as to how he regarded Indians as not-so-good research collaborators. His main reason was that the Indians keep formal relations with their colleagues even in conversations. A free exchange of ideas is not possible in official relationships. It is like doing gardening work wearing a 3-piece suit. There is indeed a mental framework for collaborative research. If I were to start academics all over again, teamwork will be high on my research agenda.

(From Daniel Bernoulli's 1743 letter to Euler) "Of my entire *Hydrodynamics*, not one iota of which do in fact I owe to my father. I am all at once robbed completely and lose this in one moment the fruits of the work of ten years. All propositions are talked from my *Hydrodynamics*, and then my father calls his writings *Hydraulics, now for the first time disclosed*, 1732, since my *Hydrodynamics* was printed only in 1738."

Daniel claimed that his father, Johann actually published his work in 1743 and predated it to 1732. There are two acts of commission; one of plagiarism and the other of distortion of facts. However, it is Daniel's side of the story. More research may be needed to find Johann's side. Incidentally, it was Johann who for a regular exchange of money let a wealthy merchant (perhaps frustrated mathematician) Marquis Hopital publish his research work under Hopital's name (of Hopital's Rule in calculus). Was Johann not prostituting research? Free riders are very common today on research papers in academe, where so much is at stake for tenure, promotion and merit.

Nevertheless, my focal point is neither plagiarism nor distortion of facts, but 'fighting' in research for supremacy and a piece of immortality. Like they say; every thing is fair in love and war. War has infinite forms. The research that the Bernoullis generated, the new areas they opened up, and the impact they had on early industrialization are humongous. Europe dominated science and mathematics through the end of the 19th century.

My conclusions on research are aptly summed up by the remarks of Andres Liljas of the 2006 Nobel Committee for Chemistry on the Americans sweeping all the 2006 science Nobel Prizes this week: *"....Besides a supportive research granting system in the US..... American universities often have more creative environment than in other countries...Creative means that people interact with each other a lot. It means you should talk with each other also, and not work as hermits, separately."*

Oct 04, 2006/June, 2014

[PS: This is significantly modified piece from a reflection that has appeared in the *Scattered Matherticles*: *Mathematical Reflections, Volume I* (2010)]

PERSONAL REMARKS

P. R. HALMOS, AS I REMEMBER

A measure of a great life is its aperiodic seismic effects over a long period of time. Call it a sheer coincidence, or an act of premonition last week, that from internet news, I learnt about Paul R. Halmos (1916 -2006) checking out from planet Earth. Halmos, a Hungarian immigrant was not a professional gypsy like his Hungarian mathematician friend, Paul Erdos (1913-1996), but he did have a craving for moving and visiting different universities (at least 18) and countries every few years. It is reflected in his intellectual vitality. Personally, since starting professional life in 1961, I have already moved ten times.

I first saw Halmos at a spring 1969 colloquium that he gave before joining Indiana University (IU) in the fall. The colloquium room was hastily changed twice to accommodate surging audience from all over the campus! Halmos lived up to the 'billing'. With uncharacteristic personality, combined with clear exposition of his research topic, he won over the math experts and **navies** alike.

Before seeing him in person, I had heard of his name from his internationally popular textbooks, in India. As a matter of fact, I was advised to bring along the Asian editions of a few books including *Measure Theory and Finite Dimensional Vector Spaces*. Halmos was a scintillating writer. For clarity and sharpness, he flexed the language and developed a snappy style that is rightfully Halmosian. **Halmosian is a newly minted 'coin' commemorating his persona.**

In fall 1969, three sections of real analysis (Roydan based) were scheduled, and everyone wanted Halmos'! I could not get into his section. He taught courses in a progressive cycle in order to groom prospective students for doctoral work. I finished real analysis with another instructor, but in fall 1970, I took Halmos' functional analysis course (only one section). His teaching also extended down to the freshmen level where he loved teaching a *Math Appreciation* course.

Halmos proclaimed his teaching approach after a famed Texan mathematician, R. L. Moore (1882-1974). But his style was all

Halmosian. Twice, during a semester, Halmos would give out 4-5 sheets of problems including standard theorems and explain how to present them in the class. He would sit at the back of the class, and interject his comments and questions, as students presented individual problems. Being conditioned to learn by lecturing for years, in India, I was an '**invariant subspace** of **Halmosian** operator'!

Of all the math instructors in my life, I rank number theorist, H. R. Gupta (1902-1988) of Panjab University (PU) Chandigarh at the top. He used to involve all the students in solving a problem during an hour. It was an assembly line approach of finishing a complex product that no one person could complete it by oneself. However, Halmos pumped and filtered the students from the point of view of their research capabilities.

A common feature between Halmos and Gupta is that they not only knew their students by name, but also called upon them during class participation. I complimented Halmos for pronouncing my last name, Bhatnagar, perfectly, as the syllable '**bh**' is phonetically very difficult for the Americans. He impressed me when he crisply pronounced another Indian name, **Srinivasan,** who taught complex analysis during 1960-61 when I was at PU. One day, in the context of a theorem, Halmos remarked how T. P. Srinivasan was known in the mathematics community for providing alternate elegant proofs of theorems recently published. Of course, in math, it is the first proof that counts for a piece of immortality. But beauty and elegance are also the attributes of a mathematical proof!

Halmos was dedicated to his classes. As a famed mathematician, he constantly received invitations from all over the world. Being conscientious of his obligations towards his courses and research, **he accepted only one invitation per semester**! On the first day of classes, he informed of the day when he would be gone. However, in his absence, the students carried on with the class business.

It reminds me of an eminent professor from India, currently on a 3-week overseas trip -including a 3-day conference in Las Vegas. I was surprised when he casually mentioned that he was away during the middle of fall semester. Neither every US professor is like Halmos, nor is every Indian professor like this visitor. Again, I recall H. R. Gupta who once declined a luncheon invitation with India's first Prime Minister, Jawahar Lal Nehru, who was on a short visit to the PU campus. Reason - the lunch hour clashed with his class hour! I have tried to live up to this benchmark.

Halmos and his wife, Virginia were married for 61 years. To the best of my knowledge, they had no children. While he was always appropriately dressed, his wife wore longish gowns with disheveled hair, the style of the pop culture and Vietnam protest era. Later on, I learnt that she was a scholar of philosophy and Latin literature! Nevertheless, they must have enriched each other's lives to stay married for six decades. Such longevity in marriage is unthinkable whether in India, or the US of today.

Halmos and his wife used to come to Hoosier Courts, a married students housing complex, where we lived on the IU campus. They carried two golden cages with a cat in each for cat-sitting done by the wife a math graduate student. He gladly paid $1.00 an hour when charges for human baby-sitting were 50 cents/hour! A story about Halmos' cats was that they were trained to play table tennis. **Creativity finds newer outlets**. H. R. Gupta had developed a hand mixing technique of making espresso coffee – normally done by a special machine!

A common trait between Halmos and I is the love for walking with a stick in a hand. For many years in the US, I walked at least 3 miles back and forth to the university. Halmos was addicted to walking. He often parked his car 2-3 miles away from his office and walked. On certain days, when he was in a mood to walk 10-12 miles, his wife used to drop him off at campus and pick him up at some other point. There were no cell phones during those days. Halmos belonged to a tradition of great Greek philosophers from the era of Socrates and Plato who mused, taught, and conversed while walking. Most of my *Reflections* also emerge and sorted out during walks. I love 2-hour walks, but never the brisk types that Halmos took.

Just like his angular athletic face, his personality was angular too. Halmos never hesitated to rub shoulders with people around him. In every group, either people loved and adored him, or just hated and felt like punching him. Here the **set** of the middles was **empty**! During my IU days, his argumentative bouts with a couple of colleagues were known to everyone.

Consequently, they never greeted or acknowledged each other in the hallways or going up and down the staircase. **A personality is equally defined by the type of enemies one makes.**

Sharp differences of opinions also showed up in his 'opposition' to faculty engaged in numerical analysis, applied mathematics and statistics. The title of his books and articles bring out his provocating personality. A couple of his catchy titles are: *Applied Mathematics is Bad Mathematics, Thrills of Abstraction* and his *Automathography: I Want to be a Mathematician*. I remember him driving a red sports car, and often wearing sport jackets, or t-shirts.

Halmos was the best PR (Public Relations) person of mathematics of his time. He was confident, witty and cocky with interviewers. Once he told the class that no matter what public mathematics lecture you give, **prove at least one theorem**. It is tough to meet this standard; nevertheless, it is not forgotten.

For years, Halmos was a **fixed point** during the January Joint Mathematics Meetings and seen surrounded by people chatting on some mathematics problems. During a class, he stressed upon joining the AMS or MAA. There were no other math organizations in the 1960s. It prompted me to join the MAA in 1974 and become a life member of the Indian Mathematical Society in 1981. However, I only started attending the meetings from 1986.

Halmos was active in both the AMS and MAA, but towards the end, he was heavily involved with the latter. **My conjecture is that Halmos must have been ticked off by the 'research brass' of the AMS**. It perhaps showed up three years ago when Halmos and his wife donated 3 million dollars to the MAA for a mathematical sciences conference center, but nothing to the AMS. Halmos, as a believer in divinely mathematics, has set a benchmark in philanthropy.

A couple of days ago, I told a friend that Halmos was a Beatle amongst mathematicians. The Beatles wrote and composed their songs, created and set the lyrics, played and sang them. They danced and smoked to their tunes. **Finite dimensional** Halmos was a captivating speaker, solid researcher, enthusiastic teacher, choosy mentor, popular author of textbooks, research monographs, provocative writer of articles on math education, and sparkling media person. Above all, he had a large heart for his friends, whether human beings, animals, or institutions. His legacy is secure and enduring.

Nov 06, 2006/June, 2014

[PS: Modified piece from a reflection that has appeared in the *Scattered Matherticles*: *Mathematical Reflections, Volume I* (2010)]

COMMENTS

1. Dr. B., Thank you for an excellent reflection. **Bob (Ain)**

2. Thanks. Interesting reading. I have not been fortunate enough to see or move at close quarters with great scientists/mathematicians. I have read quite a bit of biographies. But accounts by people who have moved with these men first hand are always interesting. Somewhat similar to hearing taped/recorded music versus attending a live concert. Regards. **RAJA**

3. Dear Satish: Happy to read. You have written really so well. Hat off!!! **BhuDev Sharma**

4. Satish - I guess I have grown out-of-touch with the broader mathematical community. This is the first I've heard of Halmos' passing. Thank you for the detailed and moving eulogy. I remember that first lecture at IU like it was yesterday. It was on the geometry of subspaces and grew into a paper called "Two Subspaces," which appeared in the *Transactions of AMS*. That talk (and eventually Halmos and lots of his other work, in general) inspired me to study operator theory seriously. Alas, I wasn't very good at it.

Halmos' influence on my mathematical life was certainly greater and longer lasting than any other's.

However, I do not remember him as "always smartly dressed." In fact, from 1971-1973, except at meetings, around Swain Hall I never saw him clothed in anything but gray cotton trousers (probably bought at Penney's), a matching shirt, high-top work shoes, a big clump of keys attached to his belt, and a baggy, wool cardigan sweater. I used to tease him that he looked like a janitor.

I think the last time I saw him up close was at an MAA section meeting in southern Oregon in the late '80s. He was irritated that they were only serving what he called "women's drinks" (like white wine and coolers) at a pre-banquet social hour. He made me take him to a restaurant nearby, where he purchased two bottles of dark beer, which he snuck back into the social in a plain brown paper bag. We talked about his automathography and *The Mathematical Intelligencer*. When I congratulated him on some prize he'd won for mathematical exposition, his only comment was, "Big deal. Who cares?" He loved the understanding and explaining and writing; and he would have done it, even without the accolades. One of the great ones! Cheers. **Larry Curnutt**

5. Excellent. Congratulations! Here is attached a copy of what we are publishing as the 'Obituary' of Halmos. Sunder was his PhD student at Indiana University. **BSY**

I wrote: Dear Yadav Ji; You are very nice with lesser words; like a '**smooth**' function that does not have the **second derivative**! Sunder's being 6-page long write-up, twice I went through it. Let me also add, that Sunder joined IU in 1973 when I finished my PhD, and left IU in 1974 for UNLV. His account is very comprehensive on Halmos' contributions in analysis and operator theory besides what the MAA has posted on its website and quotes at the end.

I have not seen any issue of the ***Ganita Bharati*** for its contents. But if it is an organ of the ***Historical Society*** that you are active in, then Sunder's is certainly over technical. Again, I have no idea of the caliber of its readership. However, I must thank you for what I wrote on Halmos, at your invitation. Who knows whether I would have written it otherwise, and if done later on, then how it may have turned out.

Halmos as I said was a multi -dimensional mathematical figure of his time. **No single write-up on him can do full justice**. Mine is holistic, intense, and personal without anybody's quotes, a style I told you before writing. **I have closely used Halmosian style and every key phrase from his books and work 'creatively'!** Above all, the coin **'Halmosian'** will be remembered for a long time than any obituary! If the journal has historical emphasis, then in fact, the Board may consider 3-4 varied eulogies on Halmos and make the journal first in the world to do it! All the best of your Thanksgiving Day in USA!

6. Satish, I finished your book. That you should find Halmos such an important mathematician and a good person is unbelievable. He has done much harm to young mathematicians, stolen the work of others and contributed nothing to mathematics but the ideas of others.

He is the Bill Gates of mathematics. He appeared as a super star in the sense that he was witty, charming and could entertain his audience very well. However, there was no substance to the man. Those who were important at IU were E. Hopf, T. Thomas, P. Masani, M. Zorn, W. Minty, but definitely not Halmos.

You think his heritage would remain, but it has not. Some of our graduate students or professors through the associate prof level have not even heard of him. He is finally a nobody as he should be. Sorry to disappoint you about your hero, but he simply doesn't fill those shoes. There are many fine Hungarian mathematicians Polya, Szegoe, Fekete, Fejer etc. Check out your lineage through me, but Halmos is the EXCEPTION. Best, **Bob Gilbert**

BENJAMIN BANNEKER AND MATHEMATICS

Who is Benjamin Banneker? I had no idea, though a week ago, on May 9, the Equity & Diversity Education Department of the CCSD (Clark County School District) celebrated the life of **Benjamin Banneker by also recognizing the achievements of the minority and other students in mathematics**! I never heard of this event before either, though it was the 6th annual function. It was held in a large ballroom of the Texas Hotel and Casino, and I was invited to speak as mathematician.

The minority communities included African Americans, Hispanics, and Native Americans. I also recognized three kids from India. In the US, athletics are high in life's priorities, but in India, it is education. In the US, people of India origin are in minority, but their achievements in science and mathematics are relatively very high. Any how, it took my breath away, as I saw little boys in 3-piece suits and girls in fine dresses. The parents were showing their pride. For some reasons, only 7 elementary, 6 middle and 4 high schools were represented! Both of my daughters, teaching in predominantly minority neighborhoods, were not even aware of this annual feature.

Benjamin Banneker lived during 1731-1806. His intellectual 'audacity' was partly due to genetic traits. According to Wikipedia, **his white grandmother came from England as an indentured labor. After the contract, she married one of her farm slaves**. A daughter repeated this family history by marrying her slave, and Benjamin was one of her children. Both mother and daughter freed all their slaves after marriages. What a trailblazing ladies were they in early 18th century! They are to be idolized no less – being ahead of emancipation in 1867.

The slaves were barred from receiving any education. However, Benjamin got its taste from her white grandmother! Though his formal education was only through the middle school, but his achievements in mathematics, science and astronomy were significant. The public education not yet born, the study of science was exclusively for the rich. His mother and grandmother provided the necessary funds for his scientific pursuits and experiments.

For a perspective, one needs to roll back 250 years, when Benjamin realized that he would never attend high school and college that are essentially the civil rights in the US today! Later in life, Benjamin was the first to raise the moral consciousness against slavery amongst the founding fathers - including Thomas Jefferson and Benjamin Franklin. They sympathized with him. However, the emancipation movement was to wait for another 100 years. Banneker was the first freed slave to be appointed on a presidential commission by George Washington.

Dorothy Strong, the founder of the Benjamin Banneker Foundation (based in Chicago) spoke about Banneker in a stirring voice and inspired the youth. Providing equal opportunities to train body and mind are the positive sides of diversity and equity, the buzz words in democratic countries from India to USA. However, if in the name of diversity and inclusion, lesser qualified minorities are hired, then it is a sad spectacle for the society. In India, the reverse discrimination has been going on for the last 30 years. **A democratic society, like a link chain, is as strong as its weakest link!** The enemy within and without can easily exploit it. Incidentally, the NFL and NBA organizations are my best models of diversity and inclusion!

May 16, 2007

COMMENTS

Satish, That is a most interesting account. I would have to point out that the US Armed Forces are probably the best model for diversity and inclusion, because of the number of people involved. Pres. Truman integrated the Armed Forces in 1948, a time when I was a reservist. (I served on active duty in the Army in 1946 - to late December 1947, and again during the Korean War in 1950-1951; so I got to see both the segregated Army and the integrated Army.) Integrated is healthier. **Dave Emerson** (Former Dean of the College of Sciences, UNLV)

Dear Satish, This is one of your best in my opinion. Raising awareness and revealing little known facts of our country are important as part of communication. Thank you, Love, **Dutchie**

Dr. Bhatnagar, I think my B. Banneker award recipients would appreciate a copy of that reflection, do you mind sending it to me as an attachment? (They were all impressed I had a course from you!) Thanks, **Aaron**

Why don't you do an article on a real mathematician and an Indian? C V Ramanujan comes to mind.... **Aniruddha**

I wrote: My focus was not on mathematics or mathematicians! The organizers have used his name to inspire kids in math. That is fine with me. I used this occasion to touch upon features like conditions of slaves, attitude of the leaders then, the role of some Whites, education in India, reverse discrimination etc.

It was a very informative piece. Thanks. **Abraham**

GONE! GO GOLBERG

Procrastination can be costly even to those who are often on the top of their daily chores. It hit me particularly when I came to know about the **Pall** finally falling on Michael Golberg (**not Goldberg**) last Thursday. For three weeks, the news of his declining health was coming from a common friend. For the last couple of years, Michael has been in and out of the hospitals and rehabilitation centers. However, his mind had remained amazingly sharp that he continued researching with his collaborators despite the failures of every major organ - both eyes, hearing, kidneys, and heart. Severe diabetes had ravaged his body - took away one leg, subsequent infections, and the disuse of the other leg.

Far from slowing down on research, Michael astonishingly credited his health problems for his exponential research productivity when I questioned him about it two years ago! He had the ability to pull anyone into his research ideas. It was astonishing. However, this time, I had decided to engage him in my problem during the next visit. Partly due to the uncertainties of his medical whereabouts, being discouraged to take phone calls in medical facilities, and my unprioritized schedules, I missed this **Last** opportunity of seeing him.

At the Jews memorial service held today, his son consented to my saying a few words on behalf of Mathematics Department. However, at the last minute, I let this honor go to a colleague who had joined UNLV in 1967 when Michael also did it. Besides, he was the Department Chairman when Michael took medical disability retirement.

In the eulogy, amongst other things, Michael's son recalled Michael being a devout Jew and supporter of the Jew causes, good cook and connoisseur of food. He also recalled satisfaction that Michael had on the completion of his last research paper in Dec 2007. It may be added, at the outset that in public Michael never ever boasted about his impressive publication records.

During 2006-07, I taught courses on the development of mathematical topics and history of mathematics. It struck me then, that there exists correlations between the organized religions and mathematics, the most organized body of knowledge. That leads to the basic tenets of religions, and their practices. The most conspicuous correlation is between Judaism and mathematics, particularly in the 20th century. The list of Jew Nobel scientists runs over nearly two out of three pages.

Point blank, I would have asked, "Michael, what is your explanation for the success of Jews in science and mathematics?", "What is there in present Judaism that is so conducive to the discoveries and breakthroughs in science and mathematics?", "Is there any ordered stimulus for science and mathematics in Jew families, temples, or synagogues?", "Did the WW II Holocaust contribute to the collective Jewish success?" etc. As a matter of fact, the Nobel Prizes in Economics, based on solid mathematical applications, have been largely won by the Jew professors.

On listening to these questions, I am sure, Michael would have smiled, adjusted his thick glasses and postured for a response. He never took a question quietly. He loved brainstorming for bringing clarity in his mind as well as that of the questioner. That is what I have **Missed out**!!

Mar 02, 2008/July, 2014

COMMENTS

Very important reflections on procrastination, prioritizing, concentration, humility!!!! Regards. **RAJA**

Satish, Thank you for passing on these beautiful reflections on my father's life. I will absolutely pass them on to my sister and uncle. **Jonathan Golberg**

PERSONAL REMARKS

ON MATHEMATICAL IMMORTALITY

"What are you going to do with this information?" I inquired while handing in a slip with my name, title of PhD thesis/dissertation (synonymous according to the **Webster**), name of thesis supervisor, university, and the award year. "I will compile it," softly responded, Harry Coonce. "But why are you doing it?" I probed for a 'purpose' behind it. "Because, no one has ever done it before!" matter-of-factly, he replied. This conversation took place nearly 20 years ago at a joint annual meeting of the AMS and MAA. Since then, I noticed Harry, a heavyset, beardy guy at a few meetings sitting behind a small table. He started on the project in the age of no laptops or e-mails!

Last week, I got an e-mail from Robert P. Gilbert, my thesis supervisor on the **Mathematical Genealogy Project (http://genealogy.math. ndsu.nodak.edu/)**. After a few clicks, I was surprised to discover that I belonged to a great mathematical lineage of Karl Friedrich Gauss! The story sequentially goes backward in time. I am (1974, Indiana University) one of the 20 PhD students (so far) of Gilbert who continues to be productive at the age of 76. Gilbert (1958, Carnegie Mellon University). Gilbert is one of the 18 students of Zeev Nehari of complex analysis fame – both as researcher and textbook author. Zeev Nehari (1915-1941-1978) is one of the 5 PhDs of Michael Fekete of Hebrew University. The middle year, in parentheses, is the PhD year.

Incidentally, two 'pioneer' Indians, Krishna Dass (1961) and Vinod Goyal (1963) also did their doctorates under Nehari's guidance. Vinod Goyal returned to India and worked at a newly started Kurukshetra University (KU), where I joined in 1965 for doing PhD in mathematical seismology under SD Chopra, Professor and Head of Math Department. Vinod returned to the US and I left KU after two years without finishing the thesis.

Besides Michael Fekete (1886-1909-1957), as one of the 16 PhD students of Leopold Fejer (1880-1902-59), University of Budapest, the others include super stars, **Paul Erdos** (1913-1996), **John Neumann** (1903-1957) and **George Polya** (1887-1985). What a mathematical ancestry so far! Further down, Leopold Fejer is one of the 17 PhDs of Hermann Schwarz

(1843-1864-1921), University of Berlin. The only other recognizable name is of **Ernst Zermelo** (1871-1953). Herman Schwarz has **two** thesis advisors, **Ernst Kummer** (1810-31-93) and **Karl Weierstrass** (1815-54 (Honorary)-97). Things get a bit confusing after this bifurcation point in the dissertation maze.

On surveying the Project data, it is evident that a stream of researchers from the US meanders into researchers from Israel, and eventually into Germany, where Berlin and Gottingen were great centers of mathematics. It may be noted that during the Nazi regime, the German universities, particularly Gottingen, were purged of the Jew professors. A few top researchers in math and physics, like Einstein, who escaped from Germany with their lives, fertilized the US universities in sciences and mathematics. The foreign PhD students, like me, working mostly with dominant Jew professors in the US, are thus invariably connected with German universities. The colonial ambitions of England, Spain, France and Italy also shaped the movement of mathematicians.

Before the 1950s, hardly any Indian came to North America for PhD. Individuals like Srinivas Ramanujan, going overseas, were anomalies. It must be stressed, that amongst the Indians going for PhD, nearly 95% are the Hindus. The organized religions play fundamental roles in the mathematical development of a society. Before India's independence in 1947, the first destination for higher studies used to be England, and then Canada. A few students did go to France, again due to India's colonial connections with France; but rarely to Italy or Spain.

The Australian and New Zealander PhD markets opened up for the Indians in the 1960s. Soviet Union also started awarding 2-3 scholarships for PhD when Peoples Friendship University was inaugurated in 1961. It was mainly for influencing the minds of the young foreign intellectuals from Africa and Asia. **Politics and education always go together**.

It must be understood that generally PhD programs are very diverse within a country and different across the countries. Moreover, they have changed a lot over recent decades. On the top, new mathematical subjects appear, disappear and reincarnate all the time. Any specialized knowledge is complex, yet fascinating to observe. Historically, the US PhD programs, emphasizing breadth in PhD, are in sharp contrast with European PhD programs focusing on depth. Having fully experienced both the PhD systems, I would unhesitatingly place a US PhD two notches above an Indian colonial PhD.

Coming back on the genealogy trail, the branch off-Kummer takes one to his advisor, Heinrich Scherk (1798-1823-1885). Kummer's **53** PhD students include **Georg Cantor** (1845-1867-1918) and **Georg Frobenius** (1849-1870-1917). Heinrich Schrek also has **two** advisors; Friedrich Bessel and Heinrich Brandes. The graph theorists and (computer) sorting algorithmists may find applications in this burgeoning genealogy project similar to the Genome mapping project in biological sciences.

Following onto the offshoot of Friedrich Bessel (1784-1810-1846), his advisor is **Gauss** (1777-1798-1855), known as prince amongst the 19th century mathematicians. Gauss' 8 PhD students also include luminaries; **Dedekind** (1831-1852-1916), **Sophie Germaine** and **Riemann** (1826-1851-1866). Sophie Germaine (1776-1831) is the first French woman to be awarded **Honorary** PhD (1930) (Number Theory) from Gottingen upon strong recommendations from Gauss. Though Germaine came from a very influential family, the intellectual climate in Paris was not favorable for women's education.

Gauss' advisor, **Johann Pfaff** (1765-1786-1825) has only two students; the other being August **Mobius** (1790-1868). This branch going for three generations (Abraham Kaestner, Christian Hausen, Johann Wichmannshausen) in the past, dead ends at Otto Mencke (1644-1707). Incidentally, some 1693 correspondence between Otto Mencke of Germany and Isaac Newton (1643-1727) of England has been established.

With nearly 120,000 digitized records, the genealogical project, so far, goes up to the middle of the 17th century. I don't think it will ever crack into the 16th century; the USA was not even born! However, the 'archaeological' work of the Project will never end. Anyone can join it sitting any where in the world. It ensures ε- piece of immortality! You can't beat this deal.

The majority of data is drawn from the US and a few European countries due to cultural similarities. Indian and Chinese mathematicians are ready in the 21st century for making a big time appearance. This ongoing project is now supported by the American Mathematical Society (AMS), The Clay Mathematics Institute, and North Dakota State University, Fargo, the current home of the Project.

Though the Project in limited in a vertical 'downward' direction, it has tremendous potential to expand laterally. For instance, Indian universities can become an offshoot of this Project like the branch roots of a humongous banyan tree. Since the 1960s, India has exploded with math PhDs. Interestingly enough, more than 60% of recent mathematics PhDs are women. **This is in sharp contrast with any other country in the world**! This observation is worthy of research for women and mathematics. Eventually, Indian PhDs will have branches, a bigger one connecting them with UK universities, and smaller ones with the US universities.

One man's labor of love and faith in the Project has created an institution involving half a dozen staffers now. It is fun to surf this site. A highlight is the number 101, the largest number of PhDs produced by one professor! This record belongs to Roger Temam who moved from Paris University to Indiana University in mid 1980's. If his professional life, so far, is spanning 50 years, then on the average, he has been producing 2 PhDs/year!

There is a joy in belonging to a network tree of the greatest mathematicians. But one PhD sub-branch seems to have terminated with me. Since the beginning of my professional career in 1961, I was clear about my place in the academe. Neither for a job, nor for a post doc, have I applied to even a single PhD granting university. Teaching different courses and designing new courses remain my passion. The number of 50 different (graduate and undergraduate) courses taught and 16 new courses and seminars designed and taught make a mathematical record of some kind. It is my piece, and place of peace and pride in professional life.

On a physical level, life is said to be the survival of the fittest, but immortality is the innermost intellectual urge in man. Yes, it is man who is more often infected with an immortality bug! The women's hunger for immortality resides on a different psychological plane. Here is a fun equation connecting the two: (urge for Immortality) = (urge for survivability)^n. In mathematical world, the number of PhD students, research papers, books etc. provide niches of immortality.

In due course, the Project would provide PhD trivia that Americans love in any walk of life. It may have information on theses, like, the shortest, the longest, the most quoted in terms of citation indices etc. The PhD trivia may unravel trivial PhD theses too! In my long professional career, I have known ghost writers of research papers, dissertations, and free riders on research papers and monographs. Once I could not help calling a particular research collaboration as 'research prostitution'! The 19th century pair of Hopitol and Bernoulli provides a classical example. It is of endemic proportion now.

In a competitive environment, men are driven to nefarious professional practices for quick name and gain. Mathematical research, by its inherent nature, trains the mind in narrow areas, but in greater depth and intellectual loneliness. In practical life, eventually, it translates into tunnel vision of mathematicians, at large.

Immortality is not agelessness, as it contradicts the irreversible bio-neurological states of body and mind. It lies in one's legacy of work, provided it is appreciated by a few. Also, no matter how great a work or deed is, there is at least one section of the society that is ignorant about it. **If an individual or even a nation is not inspired to newer heights by the greatness of one's ancestors or heritage, then it becomes burdensome in taking the present strides.**

For instance, Hindus, resorting to their ancient Vedic heritage at every new discovery in the world, have become victims of intellectual inertia. The 'Vedic Mathematics', as politically hyped in the 1990s, distracted a significant number of Hindus from creating new mathematics.

Polemics aside, this project guarantees a piece of immortality to Harry Coonce, its founder. Continuing to do what one is good at, would bring immortality sooner or later. As by product, personal contentment is already in the bag!

Apr 08, 2008/ July, 2014

COMMENTS

As you start talking to some stranger who is sitting with you on a plane, you try to find common interests, experiences etc. When it comes to knowing someone who knows someone who knows someone....whom he knows, the chain is not as long as one would imagine. I have often read that there is a theorem that proves that the average length of this chain is three!!!! Maybe you are aware of this theorem. Very interesting project. Regards. **RAJA**

Dear Professor Bhatnagar, Beautiful! Only I do not know how original it is. If you feel convinced, and write it again in a publishable form (for example, like mentioning the life period in brackets against the name of each mathematician, etc., etc.), I would like it to be considered for publication in GB. Regards, **B. S. Yadav**

Hi Satish, I enjoyed reading this. But one very minor corrections - it is Temam, not Temem. (It has taken me years to get the spelling right). **Jim Davis** Chair, IU.

Dear uncle, Thx. BTW, I am also a descendant of Gauss...so we have this in common. Do you have a website where you place all these reflections? You can easily set up a blog where you place these. **Gaurav Bhatnagar**

INVITING FACULTY INTO RESEARCH
(A reflective note to math and science faculty)

About four week ago, while introducing myself in the College meeting, I briefly indicated that my current research interest is **History of Mathematics** (HoM). As a matter of fact, I am also a charter member of **Philosophy of Mathematics**, a special interest group of the Mathematical Association of America (MAA). You may like to check the MAA website for further information. One of the reasons of my visiting University of Nizwa (UN) is to search mathematical materials and collaborate with experts in this part of the Middle East comprising Oman, Yemen and United Arab Emirates. A basic question is: **What this region has contributed in sciences and mathematics in ancient, medieval, or modern times**.

It is a tough call. My initial research efforts from the US hardly scratched the surface. Two weeks ago, I distributed a handout on HoM, to all my students in *Group Theory*, *Number Theory* and *Linear Algebra II*. I have been constantly encouraging them to work on this project, but not at the cost of the courses they are taking. To my surprise, these students take too many courses; 5-6 courses make an overload by US standards.

A reward, of advancing years in life, is that general history is unconsciously added in our repertoire of strengths. Any discipline focused history comes out as a corollary. To the best of my knowledge, few universities in the world offer PhD in HoM. However, the number of research papers, presented at the annual meetings of the MAA and AMS (American Mathematical Society), is the largest of all mathematical areas.

Hard core sciences and mathematics transcend national, religious and ethnic boundaries, but history of mathematics or sciences is heavily influenced by each one of them. To make it short, you can help me or join in this endeavor in any of following manners:

(i)   Sharing with me HoM literature or research you know, or have done it.

(ii) Encouraging my students, if they ask you any question.

(iii) Any Arabic or Persian sources pertaining to this region? The reason for a focus on this region is that it would be stretched thin, if scope is widened.

(iv) Finally, whatever comes to your mind.

I don't wish my UN association be over when I am gone after ten more weeks. In fact, this collaboration may form a basis for a conference in History of Science and Mathematics. It could be similar to an upcoming English writers' conference, *Mapping Emergent Arab and Muslim Literatures in English*, planned during April 11-12, 2009, at the UN campus.

Thanks in advance, for your cooperation.

Feb 20, 2009/July, 2014

PS July 10, 2014: This note was circulated to all math and science faculty of the UN, but not a single person responded or evinced interest. The structure of the UN is unlike any university in the US or India. Student body in 100 % Omanis, and so is the top administration. To the best of my knowledge, there was not even single Omani faculty member in the UN. The entire faculty is of the expatriates - hired from poor countries like, the Philippines, India, Pakistan, Egypt, Iraq and a few East European or former states of Soviet Union. They are very happy with the UN salaries and remain focused with saving money. People wondered when they came to know that I was a tenured professor back in the US. Research mind essentially entails a prosperous mind.

WE NEARLY MISSED EACH OTHER!

My coming to the University of Nizwa (UN) is a story, worth telling in the same spirit in which we all love to listen/read a good one. Being a writer, traveler and historian, besides fulltime mathematics professor, Middle East has held the same fascination for me as it had for the Europeans for the last 300 hundred years. It is the desert mystique, lores of Arabian nights, and adventures of Sinbad from its ancient past to the present blooming of the desert with ultra-modern structures and institutions - all due to liquid gold, the underground oil. The region continues to be a crucible of Islam and home of various empires and civilizations.

My desire has been to spend only 3-4 weeks, but the young universities of Middle East have not matured to a point of inviting scholars for shorter periods. However, one semester at the UN was worked out with the help of a friend who has been teaching in Yemen and Oman for the last 10 years. It was finalized over phone with the dean, since he had my 13-page curriculum vitae. The bottom line was, if mutually agreeable, then the contract could be extended beyond a semester.

As soon as the '**Provisional Offer of Appointment**' was received last July, I e-mailed its acceptance within stipulated three days without checking any nuts and bolts of the working conditions, or negotiating on the salary offered. From the UN's sketchy website then, the academic programs appeared to be prototypes of a typical US university. But I was just excited at the prospect of visiting Middle East, at the age of 69!

Talking of age, it did factor into the evaluation of my application. Doubts were naturally cast on my physical fitness, but they were softened. This concern is understandable, as people in this part of the world start looking forward to retirement at 50, and checking out from Planet Earth in mid 60's. My father died at 63, within five years of his retirement.

This month, I voluntarily showed copies of two different medical reports to the dean; one was required for the Omani Resident Card, but the other out of my own anxiety. Within a span of one week, I came to know of the prostrate problems of two close persons, younger than I. It compelled me to go for a complete medical check-up including blood, urine etc. I came out flying; the BP 120/70 means some thing is going on right in my life!

Coming back on the trail of the provisional offer, there were eight conditions subjected to the issuance of the final contract. One was the name of my mother; never asked before (May God rest her Soul in Peace!). Another requirement, a photograph with 'blue background' may have a reason too. Well, that is what I am here in Oman, to see how successful people in the world work and think differently.

However, the formalities suddenly screeched to a halt at two hurdles. One required, "**Employment/Experience Certificates issued by the last and all the previous employment**". I argued that I was coming for only one semester and my service record at UNLV (University of Nevada Las Vegas) alone is of 34 years. What is the point of going back to my teaching in five different places during 1961-74? The UN is a university, not an investigative arm of the Government.

In the US life, as lived for 40 years, there is no such document like service certificate. The letters from the referees besides updated CVs are sufficient. Finally, I pulled the page of my Employment Record from the CV and had it certified by my Chairman. After sending it to the UN Human Resources, I kept my fingers crossed for a few days.

The unbridled bureaucrats and clerks can do havoc to any organization. **Some historians theorize that the fall of the great British Empire was due to the burgeoning number of clerks in its vast bureaucracy bloated by the 1920's**. The famous **Parkinson's Laws** came out of this era. Nevertheless, the genius of gulf countries lies in managing every kind of import from all over the world: man power, unskilled or skilled; material, raw to finish; systems, soft or hard. Nothing is Arabic except the men at the top the ladders.

Finally, I hit a stonewall at the UN requirement that "*All the documents are to be attested by the Ministry of Foreign Affairs in your country followed by further attestation of the same by the Embassy of the Sultanate of Oman.*" The US State Department (called Foreign Affairs in other countries), currently headed by Hillary Clinton, does not involve in such academic matters.

My plea that I have been a full professor for 20 years at UNLV, a world class university, was met with a generic response, that it is the requirement of the Ministry of Higher Education (The Omanis think all are like high school **teachers**!). It really makes one wonder at the **private status** of the UN. It has to have functioning autonomy.

Here is a list of steps taken to prove that I have a PhD in Applied Mathematics from Indiana University (IU), Bloomington, a 200-year old world famous university: 1. Google my name. In 2-3 clicks, my entire academic life opens up 2. Transcript from IU directly, a standard practice in the US 3. University of Michigan, Depositary of US PhDs for nearly 200 years. 4. Mathematics Genealogy, a 25-year old website on a tree of Math PhDs going back to over 200 years. I am one of the leaves coming out of a branch from legendary German mathematician, Carl Fredrick Gauss 5. Direct transmission of my academic records from UNLV Human Resources to the UN Human Resources 6. Notarized PhD Diploma. 7. My three referees may be contacted for my credentials.

One may be wondering if I was obstinate in refusing to go through the UN attesting process. Here is a picture. The US Omani Embassy is chaotic. The calls are not received, understood, or returned; the website terrible. There is no designated person for this task. It is not guaranteed that documents Fed-Exed would stay together and returned safely. Some travel agents in Washington DC were charging over $100/page attestation by the Omani Embassy! This is unethical and blatant gouging, and I refused to submit to it.

Contemplating over the situation, I was not on academic streets, homeless, jobless, or financially desperate like 97 % of the faculty and administrators are at the UN, as I found it from my random sample, after joining it. On the top, I am teaching twice of my teaching load at UNLV for 1/2 of my UNLV salary! I, as a solitary American, outside English language teaching, may go down in the annals of the UN just for this fact alone.

Academically, horror stories are circulating over attestation process. One Indian has spent Rs 45,000 on the top of hundreds of hours in time. The front stampings deface the diplomas forever, thus making them useless for future. A rough estimate of attestation earnings by the Omani Embassy in India alone is Rs 50,000 per day! Unquestionably, the Omanis are the smartest money makers in the world. The Y-2008 profits of **OMANTEL** are nearly 700 million US \$\$! That explains why the rates of the local and international phone calls to and fro Oman are amongst the highest in the world.

After three months of nervous communications, it was already October, and I gave up on my 'dreams' of visiting the Middle East. It is well known that the best of any writer bursts out during moments of despondency. I wrote a *Reflective* note, **TIME TO ACT** to the college dean. Essentially, I told him that life is not always about money, but for richer experiences. Promptly came his very polite reply: bring your PhD diploma along, and we will have it verified by the US Embassy in Muscat. The rest is all about leaving a legacy.

March 28, 2009

COMMENTS

Your experience is so much in contrast to what my friend experienced in Qatar. My friend taught at Cornell University in Qatar for 3 years. The teaching load was one lecture and one lab. His salary was considerably higher than what he made here. On top of it, the benefits were excellent. The difference was that he was hired in America, not in Oman, as is the case with you. There was no paperwork where he had to deal with the foreign government. Most of the work was done by Cornell University. ....I enjoy reading your reflections. You are a lot better now as a writer. Best wishes, **Alok Kumar**

2. Dear Bhatnagar Sahib: You forgot that in the East, it is not what you know, but who you know that counts! Best of wonderings in the East! Send more reflections on your experiences! Looking forward to your safe return to the good, old USA! Very sincerely, **AR Bhatia**

3. These are standard procedures in the Middle East as far as employing expatriates are concerned. It is not just Oman, but also other GCC countries which follow similar (and in some cases even stricter) procedures. It's no good comparing it to the US because this is a different continent, and an absolutely different culture!!

Apologies in advance for a deeply personal opinion, but US is the root cause of most of the troubles in this world, be it the financial crisis of today or the raging wars in various parts of the world! Benchmarking the practices followed in the US is probably not appropriate! No offence intended - the above was just an honest opinion!! Warm regards **Amit Bhatnagar**

RETURN OF A KIND!

"Dear Dr Bhatnagar, I'm not sure if you would remember me. I was one of your students in Shah Alam, Malaysia in 1992. I went to Tucson & did BA in Math from the Univ. of Arizona, graduated in 1994. I came back & taught in Malaysia for a while, and got married, then went to do M.Ed. at the Univ. of Ottawa, graduated in 2003. I'm a PhD candidate at The University of Melbourne. My study is on secondary mathematics teachers' assessment practice in Victoria, Australia. Having met some of the great math teachers here, I reflect back on those math teachers who made a difference in my life, and of them is you.

I'm writing to you just let you know that. Enjoy your life, great to know that you are now in LV, & have two grand kids! I am now a mother of four growing sons. Life is full of wonders.

Kind regards; Rohani Mohamad IU-9".

It makes my day whenever, such a note or e-mail suddenly crops up. Teaching is a different profession from the point of view of 'customer satisfaction' (short or long term?). One does talk about teaching in an era of public accountability and quality assurance. No matter how hard I try to reach each and every student, say, by knowing their names, majors and connecting them out individually, there is always at least one student who remains turned off by my style. It took me years to reconcile myself with this aspect of human nature. It is no different from what generally is observed in life; no one food, no one movie, no one song is equally liked by everyone; nor is uniform the impact of a book, or effectiveness of a medicine. Human individuality manifests itself in extreme outcomes.

Since this e-mail flashed last night, I have been trying to re-construct Rohani's persona on a premise, if she was impacted by my persona, then its converse must hold to some extent. With all the mental powers harnessed for it, her 1992-image is of a fair complexioned, healthy girl from a kampong. She was in physical contrast with most Malay youths who were darkish and thin.

With the exception of one male student known as M, the good math students were all females. Yes, one girl was admitted to the University of Arizona (Tucson). Also, I vividly remember, her message conveyed to me through her junior friend, that she had gotten nearly straight A's in the very first semester. It had filled me with joy at her thoughtfulness, and pride at her performance.

When I put myself in the role of a student, then the memorable tidbits that I had with some teachers were not the same, as they had of mine. In schools and colleges the influence runs on a principle of flux moving from 'higher' potential to 'lower' potential. It is totally one sided in elementary schools, but tapers off as one progressively moves towards the doctoral level.

Rohini, thank you for making my day! It is credit worthy that math has stayed with you despite big changes in your marital life. You remind me of Mary Rudin, who as 'housewife' proved great theorems in point-set topology while doing laundry, watching her kids playing out, and working in the kitchen. You may Google her name. Oh, regarding my website, you mentioned, I don't even remember it, as I still live in the dark caves of electronica. An update: six grandkids now!

June 12, 2009

COMMENTS

This is indeed an e-mail to treasure...**Archna**

Dear Prof Bhatnagar, Congratulations on the respect and regard that you received. The trouble arises when the student thinks he is only a consumer whereas he is truly a product as well (read in FOCUS of MAA or elsewhere!) Best regards, **Ajit Iqbal Singh**

Hi Satish: Congratulations! Sure, we do as teachers make a great impact on our students. Your 'reflection' hits the nail on the heath, to use a cliché' for a change. Teaching is a fulfilling profession. **Moorty**

Dear Prof Satish, Your reflections are inspiring and often infuse new spirit in me. I carried them with me when I left for holiday and were part of my breakfast. Now I am back to Nizwa experiencing the grilling heat of the summer. I will send you my reflections on the holiday when I reach an equilibrium point. Cheers, **Ahmed Yagi**

GIANTS IN THE CLASSROOM!

The longer the time passes, greater is the degree of amazement of a certain experience, as to how come it was not realized sooner. It is about something that is always within your reach, of interest, 'on the way to work or home', etc. It is simply taken for granted, or just assumed -'known'; for whatever it is, or what it is going to be.

A week ago, I declared, that I was going to bring the *Principia Mathematica* (PM) of Russell and Whitehead in my graduate class on *Survey of Mathematical problems II* (MAT 712). The idea is to test run an in-class exercise on great mathematical works. Naturally, it became incumbent upon me to check them out from the library, and have a good look at them myself first. I don't recall the last time I saw them. It is certainly out of question before 1968, when India was my 'permanent' home. However, on the way to the library, I also added the immortal classic in all sciences, namely, Newton's *Principia* abbreviated for *Philosophiæ Naturalis Principia Mathematica* meaning - Mathematical Principles of Natural Philosophy (MNP).

It was a bit painful to hand-carry four volumes 200 yards to my office. But it was nothing as compared with intellectual pain encountered in trying to understand them. Any work of literature, history, and social studies has many passages that one can reasonably understand in bits and bites while flipping the pages. Mathematical volumes defy this treatment; almost telling the handlers: Don't touch; Stay away; It is 24,000 volts, and so on!

However, I set myself in the office on a quiet Friday afternoon, and started tackling the PM first. The entire work, in three massive volumes, covers nearly 2000 pages. Russell (1872-1970) tried to remove paradoxes and lay new foundations of mathematics, essentially based on logic behind new theories of Types and Descriptions. By the age of 25 in 1897, he had it crystallized in his mind, and started writing it in the fall of 1900.

It is in the process of writing that one discovers holes in the wholes. The more he tried to plug and solidify them, the more it went out of control.

He and Whitehead (1861-1947) tried to prove 'everything' in precise detail. The line numbering system alone on every page is four times more complex than that of a modern legal document. Each main result has up to three digital layers of referencing notation justifying each step in it. Fifty years later, Bourbaki, a secretive group of French mathematicians, must have been inspired by the PM to further the analytic rigor in collegiate mathematics. Bourbaki published several volumes.

A cursory examination of the PM was a humbling experience. I could not make out anything from its complex symbols and notations given in page after page including its Preface. It is an incredible work of mathematics, never to be duplicated in its detail and voluminousness. It is also a tribute to human perseverance, done without any technological help. In fairness to my math background, I never had a course in foundations of mathematics during my stay at Indiana University. In India, mathematical logic was never heard of during my college days in the 1950s.

On the other hand, the MNP is not daunting at all. Its geometry, mathematical physics, and experimental data are within grasp of good undergraduates. The beauty of great ideas lies in their elementary nature. For example, a physics faculty member, at UNLV, teaches Special Theory of Relativity based on precalculus. The MPN covers only 541 pages. It has been stirring the minds of the curious students and professionals for centuries! In the entire translation, the only non-Newton piece is a 45-line poem by his friend, benefactor, and science admirer, Edmund Halley (of Halley Comet fame). The last six lines of the poem, including one, 'no closer to the gods can any mortal rise', are incredibly prophetic on the impact of Newton's discoveries on mankind. **It takes a Newton to admire the Newton!**

Newton did change the course of mankind and brought a true paradigm shift in intellectual thinking. He wrote ***Principia*** in Latin, the elite language of scholarship at that time. He did write in English too. I had a hand-written copy of his paper on alchemy – a subject that he also pursued. It was brought by my student from the Library of Congress. Latin occupied the same place amongst European scholars that Sanskrit did with Indians before 1000 AD. The 3rd edition of the ***Principia*** was published in 1726 - a year before Newton died. Its first English translation by Andrew Motte, done in 1728, served the English-speaking world for over 250 years.

Cohen (1914-2003), first Harvard PhD in History of Science and subsequently its chair professor, undertook a 'user friendly' translation of the MNP, in collaboration with Whitman, a Latinist. The entire translation project matched the greatness of the MNP. Cohen and Whitman studied the MNP and Newton's other works so as to gain insights into his mind. They went a step further than Motte did, as they prepared nearly 400-page manual on *How to Study the MNP*! The translation is a scholarly work as well as that of labor of love, as it took 15 solid years. Of course, the translators had access to the resources of Harvard University topped by a grant from the NSF.

Here is a publication trivia. The publication cost of the PM was 600 Lbs. - 300 Lbs. borne by Cambridge, 200 Lbs. by the Royal Council, 50 Lbs. each were taken by Russell and Whitehead from their pockets! I don't know how many copies were published. On a non-trivial note: during my lectures, I often inject a question on the necessary and sufficient conditions on the development of mathematics in a society or a nation. Newton's *Principia* dawned an era of rational philosophy in Western Europe, a gentle departure from the influence of dogmatic Church. It laid the foundations of modern science. **The PM, in size and detail, matches the zenith of the British Empire**. Its decline started in 1920s, and Russell's political stands may have accelerated it. The incubation and translation period of the MNP also reflects the acme of American Civilization, though the jury of historians is still out on it.

Some tangents on non-mathematical sides of the Greats: "So I persisted, and in the end the work was finished, but my intellect never quite recovered from the strain. I have been ever since definitely less capable of dealing with difficult abstractions than I was before," wrote Russell in his 3-volume autobiography (1967-69) on 17 years devoted on the PM. He quit math after 1914. He became pacifist and passionate about many international issues such as, peace, disarmament, etc. His candor, originality and force of ideas made him the most sought-after speaker in the world. Apart from numerous honors, he won the 1950-Nobel Prize in literature; deserved one for peace too. On Newton, Cohen wrote: "He was not a pleasant man. He had no intimate friends, no relationships with women. It is said that he died as he was born – a virgin." In contrast, Russell had openly passionate relations with many women including 3-4 as his wives.

Oct 30, 2009

COMMENTS

Thanks Prof. Bhatnagar for sharing your interesting academic anecdotes. **Ajit Iqbal Singh**

Once again, excellent piece! **Aaron**

DARK SPOTS ON THE MOONS (PART II)

[Note: This piece is extracted from a 21-page compilation - an ongoing project on history of mathematics. I have essentially pooled 200-300 word submissions of the students in several courses who earn extra credits for doing it. It is not mandatory, but most students do it. For the purpose of illustration and space, only a few lives have been chosen. There is no order of any kind except the way they were received. Also, I have not changed much of students' language.]

Sometimes, one is surprised at how a new cooking recipe would turns out; a line of research shed light on a different problem; a hike opens out to exhilarating new vistas. A month ago, an article in the *Focus*, a newsletter of the MAA prompted me to examine the lives of legendary mathematicians from an 'opposite' end. As I enter into my 70s, it is becoming increasingly clear that every life approaches its zero sum (a mathematical limit!) in terms of its highs and lows.

Archimedes (c 287 BC – c 212)

The students did not find anything in the dark side of this Greek immortal. This it is an open question for research on is life.

Ludwig Boltzmann (1844- 1906)

During the course of his life the common perception of Boltzmann was that he suffered from an undiagnosed bipolar disorder as well as probably Manic Depression. Like Boltzmann, the concepts of the agitated chaos of the atom in which he so firmly supported, Boltzmann's mind can similarly be said to be as equally restless.

Boltzmann can be pointed out to be a prime example of an early feminist - in that he not only allowed his wife to take interest in his research but rather encouraged it; which at the time was considered to be very inappropriate. The long periods of depression that set upon him - had caused him to attempt suicide numerous times, and eventually claimed his life around 1906.

Georg Cantor (1845-1918)

Cantor received his doctorate at the age of 22. Leopold Kronecker, once one of Cantor's teachers, grew to oppose his work, and Cantor never recovered from the affront. He never submitted his work to the professional journal associated with Kronecker.

Cantor experienced his first recorded bout of depression in May of 1884. It was during this time that Cantor began intense study of Elizabethan literature, attempting to prove that Francis Bacon wrote Shakespeare's plays. Cantor battled depression, and spent the last 20 years of his life in and out of sanatoriums. He retired in 1913 and spent his final years suffering from malnutrition and poverty, died in 1918.

Rene Descartes (1596-1650)

I think, therefore I am, or.... I am thinking, therefore I exist. This is just one of the many ideas that were introduced by Rene Descartes, who is regarded as a founder of modern philosophy.

Around 1630, Descartes had a falling out with Beeckman, whom he had accused of plagiarizing some of his ideas. Descartes had a daughter named Francine with a servant girl, unfortunately just five years later, Francine passed away. After his daughters' death, he moved frequently throughout the Netherlands. Despite an unstable homestead, he wrote most of his major work during his 20 plus years there. Descartes was condemned himself in 1643 for his Cartesian philosophy by the University of Utrecht where he had taught. His interest in metaphysics put forth *Meditation,* which introduced ideas about knowledge, perception, and God.

Throughout Descartes life, he had established relationships and made acquaintances with several high status figures of his time, one of those elite being Queen Christina of Sweden, and moved to Sweden in 1649. Descartes gave Queen Christina lessons in philosophy, which apparently started at five a.m. and lasted for hours.

Albert Einstein (1878- 1950)

One quite shocking proof of this fact is that Albert Einstein was not decent outside the office. For instance, he committed adultery. He is known to have never been a faithful husband, often times falling in love with someone immediately after exchanging vows. He once told an interviewer, "All marriages are dangerous. Marriage is the unsuccessful attempt to make something lasting out of an incident." (http://ioframe. com/interesting_facts_about_albert_einstein). In addition, Einstein had an illegitimate child whom he never saw and whose fate remained unknown..." (http://www.neatorama.com/2007/03/26/10-strange-facts-about-einstein/).

Paul Erdös (1913-1996)

In the process of collaborating with so many other mathematicians, his lifestyle became very nomadic and personally solitary. One of the unfortunate results of this was that Erdös never had a family or any personal relationships that would yield any children. He would routinely refer to God as "the Supreme Fascist", women as "bosses", and men as "slaves." In the realm of mathematics, he would refer to his colleagues that left the field of mathematics as "dead". Yet, he would refer to those who died as "left."

He is noted to have been addicted to amphetamines such as Benzedrine or Ritalin. The drugs allowed him to devote as much as nineteen to twenty hours a day to math problems. If he was not taking caffeine pills, he was known to be drinking copious amounts of coffee. According to some friends, he was fond of saying "A mathematician is a machine for turning coffee into theorems." This may have attributed to his gaunt and frail disposition and even contributed to his death by heart attack while attempting to do what he loved best: solve math problems.

An interesting fact is he could fly all over the world without jet lag. In fact, he was very stupid when it comes to real life problems. He did not even know how to work a washing machine or to cook. He was a drug addict. He was so skinny that had cadaverous look. However, his productivity went down without these drugs. He also had a type of

140

mysophobia that he had to wash his hands 50 times a day. The most prolific mathematician in the history in fact lived a pitiful life behind the great fame. He gave up potentially luxurious life over his passion on mathematics.

Evariste Galois (1811- 1832)

Galois excelled in many areas of school, but the routine quickly became a bore and he turned to mathematics for entertainment. Immediately following his father's death, Galois once again attempted to enter Ecole Polytechnique, and once again met with failure. With his temperament surely rattled at the rise of such adversity, Galois entered into a life filled with political activism.

The 14th of July was Bastille Day and Galois was arrested again. While Galois was in prison, like his father, he attempted suicide by stabbing himself with a dagger but was saved by the inmates.

He apparently fell in love with Stephanie, the daughter of resident physician in prison. Galois fought a duel on 30th of May---certainly linked with Stephanie. Galois was wounded in the duel and left by the opponent until found by a peasant. He died in the hospital on May 31st.

Carl Friedrich Gauss (1777- 1855)

Gauss grew up with uneducated family with poor environment but he succeeded in certain ways. He had no fellow mathematical collaborators and worked alone for most of his life. Gauss was known as "glacially cold person. In 1809, Gauss went into depression due to the early death of his first wife, Johanna Osthoff, followed by the death of one of his child, Louis. He never fully recovered from these deaths. He then married Minna, however, this marriage was not a very happy one. Minna became a permanent invalid after three childbirths.

Gauss always remained almost equally interested in philology, and wrote well in German and Latin. He enjoyed reading historical works and preferred novels with a happy ending. Gauss was a perfectionist. He did not publish works that he thought would be criticized by his peers.

Sophie Germaine (1776-1831)

Sophie German began to study the math books in her father's library. It was believed that because of her gender and family background, she should not be educated in mathematics. Her parents initial disapproval did not hinder her desire to study, they attempted to take her clothes and candles away so she could no longer study at night. These attempts failed because she would smuggle candles and wrap her naked body in a quilt to learn about mathematics.

Sophie's great feats in mathematics were ignored by mathematical community or she was never taken seriously. She suffered many disappoints. Even on her death certificate, she was not labeled as a mathematician, but as a renter of properties. In the two years prior to her death due to breast cancer, she would send letters where she talked about why she was suffering so much. She did not see the purpose of living anymore because her friends were not talking to her and famous mentors did not communicate with her.

May, 2010

CONNECTING THE DOTS

An innocuous exchange of thoughts has created a sort of mushroom of mathematical 'nuclear' cloud in my mind. After reading a brief article in a mathematics newsletter, on the famous Russian woman mathematician, Sonya Kowaleski (1850 -1891), I commented that she had died of tuberculosis - not of any generic fever, influenza, or flu. Immediately, the following response came from Mahavir Vasavada, the founder of this monthly newsletter:

"Thanks for the information about the cause of Kowaleski's death. My information came from two sources:

1. *Men of Mathematics by E. T. Bell: p.429.*
2. *Love and Mathematics (Biography of Sonya Kowalski) by Pelageya Kochina: p. 281."*

Referencing is a western intellectual tradition going back to the turn of the European renaissance. Historically, such references were only meant for mathematics and sciences, where references to mathematical theorems *and* scientific laws and principles are more or less undisputed. It saved a lot writing paper, especially in the 13th century, when paper was as expensive as gold! However, the bibliographies in humanities and social studies are relatively very recent. They basically competed with math and sciences for scholarly prestige. Essentially, it is a battle of theorems and laws of nature with social theories and individual opinions.

Referencing took a dive, particularly in the US academe, when *'publish or perish'* became a *mantra* for success in professional life. In a nutshell, if two historians or economists agree on a theory, then one is a free loader or a rider. In the so-called Tier I universities, senior professors exercise intellectual controls - like the mafia bosses do, in terms of demanding loyalty to their theories. As a rule, longer a bibliography, the less trustworthy the work is - in my opinion - modulo a set of measure of zero! (I love this phrase from real analysis.)

The problems in history of mathematics are unique. Historians, particularly, taught in Indian curricula, have no in depth knowledge of mathematics. Any attempt on history of mathematics is very shallow. On the other hand, most research mathematicians have no mindset for history, as the two disciplines are poles apart. It is a paradox on how to train or find historians of mathematics in India! PhD programs in history of mathematics are really unheard. Let me hear of one!

However, during mathematics conferences, the growing number of sessions on History of Mathematics indicates that eventually senior mathematics professors and non-research faculty gravitate towards history of mathematics. In line with GH Hardy's, classic book, *A Mathematician's Apology* (1940), some 'oldie' professors continue to pretend doing research. They remind me of the seniors whose erectile dysfunction is even beyond Viagra!

History of mathematics is not disconnected from history of people and politics. Again, Indian graduates of mathematics and sciences are not exposed to humanities and arts. How can they be credible historians? I have regards for Mahavir Vasavada whom I met last Jan in a history of math conference. He and his wife, both retired, are the only husband and wife team in India who has been popularizing math through a monthly newsletter and meetings.

A question may be raised as to my credentials in history. Yes, my American PhD is in mathematics, but I made a conscious decision not to become a research mathematician. In 1974, when I joined UNLV, it had no PhD program in mathematics. I flourished in teaching different courses. Needless to say, I did minimal research to earn tenure and promotions to the rank of full professor. Honestly, it was not easy; so is the story of choosing any untrodden path in life.

I always nourished myself so as to live multi-dimensionally - be it at the intellectual, physical or spiritual level. In the 7th grade, I questioned the nonsense theories on the *'Origin of Aryans'*, as the natives of India are called. During high school history, as a Hindu, I hated the glorification of the Muslim and British rulers of the Hindus. After college studies, I gradually went back to history unattached with any school of thought.

Mathematics gives objectivity and compactness of expression. My interest in history of mathematics comes from mathematics as much as from history in general.

Coming back to the cause of Sonya's death, the term influenza is parochial. It became commonplace only recently. Flu is its abbreviation. While growing up in Punjab, through the 1960s, fever (*taap* or *bukhaar*) was commonly used. The word 'flu' was unheard in India before the 1980s. In the US, every year, flu has a different technical name depending on the mutation of its virus. TB was not clinically understood at the level of the commoners before WW II. However, a reference to lungs and her long confinement with flu confirms that she died of TB.

June 15, 2012

PS:
I have researched into how TB came to India, where it was unheard before the 18th century. It is included in my book, *Vectors in History: Main Foci; India and USA, Vol I*, 2010 (Trafford).

COMMENTS

Keep it up, Satish Ji. **Prem**

One statement, highlighted below, is not accurate. There was an epidemic of flu in Delhi in early or mid-fifties. My mother got one. That was the first time I realized that "Flu" was an abbrev of influenza. Various distinctions of flu viruses are a relatively recent phenomenon even in US. **Subhash Saxena**

PERSONAL REMARKS

SECTION III

INDIAN SPICES

GENERAL MATHEMATICS IN THE VEDAS

Summary

[The Vedas are not the books of mathematics, yet mathematics is scattered all over in them. Some Vedic scholars maintain that mathematics, being also a language of Nature, is indispensable for the total understanding of the Vedas. Shankaracharya Bharatikrsna Tirath said: True Realization is Actual Visualization, and hence has to be mathematical too. Through yogic powers, the state of awareness in which consciousness knows in totality, he extracted sixteen *sutras* (aphorisms) from the Atharva Veda. This paper is inspired by his work, and it raises the questions on the existence of collegiate mathematics in the Vedas.]

Background

In January 1983, I delivered seven lectures during a mini course [1] on Vedic Mathematics based on the sixteen **sutras** (aphorisms), as claimed to have been discovered by Shankaracharya Bharatikrsna Tirath. Without giving any proofs, he has only explained the applications of the *sutras* with typical examples in his book, Vedic Mathematics [8]. The response of the participants was very positive, as in a short time, they were themselves able to do fast mental calculations. Gradually, I was drawn toward its deeper investigations, and spent the year, 1986-87 in India while on sabbatical leave. The purpose was to learn about the place and nature of mathematics in the Vedas, and also establish some contacts with Vedic scholars in the common area of mathematics and Hindu scriptures.

Importance of Sanskrit in Vedic Mathematics

In India, there is no established university department devoted to Vedic Mathematics. A person in this area has to be well conversant with both mathematics and Sanskrit. Unfortunately, such a combination has been non-existent in the Indian curricula for centuries mainly due to foreign subjugation of the Hindus. Currently, those who study Sanskrit in schools and colleges generally are not permitted to study mathematics and sciences, and vice-versa. It has been going one since Lord Macaulay brought the 'education reform' in India around 1830. Several generations have passed without any great scholars who have produced fundamental work in science and Sanskrit. That is why today, the greatest opposition and even ridicule to mathematics and sciences in the Vedas come from the academicians. This year, the Government of India has set up a National Foundation for Vedic Studies to relate the Vedas, Upvedas and Vedangas to modern sciences.

Sanskrit did not develop merely a language as connoted today. Its complex, but very structured grammar, reached its zenith in the time of Pannini (c 520 BC – 460 BC), who flourished in the then Buddhist region called Afghanistan since conquered by Islam in the 9th century. So much so, that Sanskrit was considered a divine language of the ***devas*** (gods). It implication is that its functions are far and above the usages of ordinary languages. Recently, the power of its structure has found applications in artificial intelligence.

Investigative Approach

In the absence of any institution where one could get the benefit of scholarly interaction and library research, I followed India's time honored tradition of the search of scholars by a word of mouth. During these travels, I have met some Vedic scholars who are pursing Vedic knowledge as the ultimate truth of life [5, 6, 9]. For them, the key is the totality of knowledge - not its fragments like mathematics, physics, or biology etc. That was a kind of setback in my type of so-called 'modern' training. Discussion on specific topics of mathematics often 'digressed' into metaphysics, or philosophy of mathematics. Narinder Puri [4] offers popular courses and presentations on the aspect of fast mental calculations based on the sutras while working full time as a professor of civil engineering in University of Roorkee. Last spring, Maharishi Mahesh Yogi sponsored Puri on a successful worldwide tour of lecture-demonstration on Vedic Mathematics.

Seven Open Questions on Mathematics in the Vedas

1. What are the precise sources of the sixteen *sutras* in the *Vedas*? Shankaracharya has simply mentioned that they are to be found

in the recension of the Atharva Veda, but gave no specific clues of any particular *mantras*! The genesis of his book, *Vedic Mathematics*, as explained by Manjula Trivedi [8], is quite intriguing.

Shankaracharya himself added that the *sutras* were discovered after his eight years of *tapa* (meditation with austerities) in the forests of Sringeri. It then raises a question on the relationship of yogic powers with specific goals. Maharishi Mahesh Yogi said: "Completely identified in transcendental consciousness with full potential of natural law, the human mind is a field of all possibilities, spontaneously functioning in harmony with all the laws of nature, and able to accomplish anything." A paper [2] on this theme was presented during the First World Yoga Conference, held in New Delhi during December 29-30, 1986.

2.   Besides the sixteen *sutras* as mentioned in the *Vedic Mathematics*, are they any other *sutras*? There are some indications and suggestions for a seventeenth one, but none has been found yet.

3.   Is presence of mathematics in the Vedas symbolic, or does it have any topical continuity? There are several *mantras* in group forms – like, for example, numbers in arithmetic progressions. See [7]. If these *mantras* have no other meaning, then why these *mantras* should have survived for millennia?

4.   What are the bases of algebra in the Vedas, as math taught in the schools and colleges? It is the higher level of abstraction which distinguishes algebra from arithmetic. Vedic numerology provides one example, but its origin in the Vedas has not been fully established.

5.   In the Vedas, there are quite a few *mantras* on geometrical properties [7], but are there notions deep enough - like a parallel postulate of Euclidean Geometry?

6.   The concept of infinity is the cornerstone of calculus. There are all kinds of epithets to describe largeness and smallness, but nothing in the Vedas has been explicitly found, which is close to infinity and infinitesimals of modern calculus.

7.   There are several mantras which have two to four entirely distinct sets of meanings - including one completely mathematical. Assuming a particular *mantra* is the creation of one mind, then

how a mathematical formulation fits with other non-mathematical formulations which are often ritualistic and metaphysical? Does it mean in some meta state of mind, mathematical symbols and other symbols have common bridges? This also ties with Maharishi's Unified Field which is described unbounded, undifferentiated and unified field underlying all activity in nature. See [4].

A complementary question is that with today's specialized approach in education, can one ever compose a *mantra* with such diverse and multiple meanings? That is where perhaps the artificial intelligence of powerful modern computers may come in and help out!

Present Work

Superficial reference to mathematics and sciences in the Vedas can be found in several books and commentaries on the Vedas. Vaidya Nath Shastri [7] has cataloged some *Vedic Mantras* according to traditional branches of mathematics and sciences. In fairness, these efforts do not establish the Vedas a repository of scientific knowledge worthy of being called divine revelation. To some extent, they undermine the Vedas - if that is all they contain.

Conclusion

Maharishi Mahesh Yogi has taken a bold and bigger step. By inviting some of the top scientists and Vedic scholars together, his organizations have tried to prove that all great modern theories in physical sciences can be traced to some seed principles contained in specific mantras in the Vedas. Such a connection is more of a symbolic interpretation than an actual derivation from the viewpoint of a modern scientist. This approach will be validated, if a physical law or some property of life or matter is first derived from the Vedas before it is actually discovered by the modern researches! Unfortunately, the current approach is just the opposite. Certainly, it is time to increase awareness and organize efforts for fundamental researches in the Vedas from the perspective of scientists and mathematicians.

References

1. Bhatnagar, Satish C. *Fast Mathematics*, Lecture Notes, UNLV Mathematical sciences, Jan. 1983.
2. Bhatnagar, Satish C. *Mathematics in the Vedas and the Yoga*, 14-15, Yoga Mandir, July 1987
3. Maharishi Mahesh Yogi, *Enlightenment and Invincibility*, Rheinweiler, West Germany: MERU Press, 1978.
4. Puri, Narinder, Ancient Vedic Mathematics and Spiritual Study Group, Civil Lines, Roorkee, India.
5. Sharma, Kashi Ram, Regiment Bazaar, Ambala Cantt, Haryana, India.
6. Sharma, Krshen Dutt, Honian Street, Jagadhari, Haryana, India.
7. Shastri, Vaidya Nath, *Sciences in the Vedas,* Sarvadeshik Arya Pratinidhi Sabha, New Delhi, 1970
8. Tirath. Bharatikrsna, *Vedic Mathematics*, Motilal Benarsidas, New Delhi, 1965.
9. Upadhaya, Shiva Kant, Arya Samaj Mandir, Rajinder Nagar, New Delhi.

June, 1988

PS: April, 2014

This paper was presented at a conference, and also widely circulated privately during an era of no e-mails.

MATHEMATICS AND COLONIZATION

Research on any topic does not follow a straight line course. At times, it takes the path of a bullet ricocheting off one object to the other before getting embedded in a soft target. This semester, *History of Mathematics* (MAT 714) has been a new experience. It is far different from teaching a typical math course. There are no hard core math problems involving analysis, algebra, or calculus. It is all about mathematics, mathematicians, universities, and cultures that created, nourished, starved, or ignored mathematics.

An amazing realization that came out of this course is that whereas mathematics can be taught in isolation from science, engineering, and of course, from arts, humanities and social studies, but not its history. History is always rooted in the people, and people interact with people from every walk of life. **Thus history of any facet of mathematics can take one as far as one lets the imagination fly!**

Well, it happened yesterday, when I noted that the British mathematician, Grace Chisholm (1868-1944) had a connection with India through her mathematician husband, William Young (1863-1942). Incidentally, Chisholm was turned off by Cambridge Math Dept, as 'ruled' by Cayley (1821-95), though she had passed the Senior Cambridge exam - the most prestigious Cambridge Tripos exam, and later on, the Final Honors exam of Oxford too. She was tutored for the Cambridge exam by her would-be husband, Young. Despite her sterling credentials, she was not 'accepted' for PhD into Cambridge, and even by Cornell of the USA! She then decided to go to Germany's Gottingen University where its Math Dept was headed by Felix Klein (1849-1925). She received PhD under Klein in 1895, and married Young in 1896.

Both husband and wife collaborated on several math books and research papers. The exact reasons being not investigated, as to why Young sailed to India in 1913! The financial tightness along with travel adventures may have played it out. There are indications that when Cambridge cut his tutoring duties, the couple moved to Germany and Italy. For raising their five kids, they must have figured it out that overseas employment, for a few years particularly in the British colonies, was financially wiser.

According to the government records, besides comparable British salaries in India, the British received several 'hardship' perks. In colonies, they had little to spend on, and hence the savings were astronomical. However, seminal research suffered. This practice still goes on - Americans go overseas with the benefits of modern income tax breaks. I too reaped these benefits during my 2+ years stay in Malaysia as visiting professor of Indiana University.

It is not clear whether Grace accompanied her husband to India. Most likely, she may have stayed back for the sake of children's college education that hardly existed in India then. Young joined Calcutta University as the first Hardinge Professor of Pure Math (1913-17). Calcutta University was set up in eastern India in 1857, the year Indian Mutiny was raging in the north around the major cities of Delhi, Jhansi, Meerut, Kanpur and Lucknow. The British quelled the disorganized freedom fighters with the support of their loyal Indian princes. Bengal with its state capital, Calcutta, situated in east India, had fallen to the **East India Company** of England in 1757, after the Battle of Plassey. My state of Punjab, in the northwest India, was annexed to British India in 1849 - right after the collapse of the Sikh Empire.

According to the *Wikipedia*, it is intriguing to note that Young was holding two part-time positions during this period! It is possible that he may not have continuously stayed in Calcutta due to the nature of his duties and/or high humidity during the times of no electricity. Also, the part-time faculty positions exist only when the unemployment rate is either very high, or very low.

Calcutta University was established purely as an affiliated university offering no actual instruction. It was only the examining and degree granting authority over its affiliated colleges spread all across north India. Undergraduate teaching was carried out in the colleges, and it still goes on. Starting from 1904, the University gradually added teaching to its supervisory functions, and offered postgraduate (called graduate in the US) classes at the Presidency College. In 1917, the University set up some centralized departments. The growth of the University accelerated under the vice chancellorship of Sir Ashutosh Mukherji (1906-24). It is likely that Mukherji, being himself a distinguished student of mathematics and physics, hired Young to set up the first academic department.

It was the policy of the British master plan for the colonies to impart education only in English literature, history, economics, English philosophy, political science and psychology. The native languages were absolutely discouraged. Also, science was kept at a distance. Mathematics falls in-between science and humanities. For instance, in my hometown, Bathinda, with then population of 65,000 in the 1950s, the science classes were started for the first two years in 1955 - eight years after India's independence in 1947! The English language and humanities molded the Indians to think British. The British thought science, being expensive, was also beyond the superstitious Indian minds.

Young was not the only reputed mathematician to visit India before independence in 1947. Andre Weil (1906-98), later of Princeton's Institute of Advanced Studies, spent two years (1930-32) in Aligarh Muslim University (AMU) which opened up in 1920. Weil tried to create research environment in a department of four (non-PhDs). Strangely enough, Weil was hired as professor of French Language and Literature! He left in frustration due to bureaucratic reasons and institutional politics of religion. Incidentally, Weil was recommended by his professor to the AMU Vice Chancellor during his Paris visit.

In the modern context, it may be added that American education is the hottest commercial commodity in the world today - from its elementary to tertiary education. What the Church spearheaded for the European colonizers from the 17th through 19th century, education has been doing it for the US since the middle of the 20th century. During my three years in Malaysia and Far East with Indiana University, I was surprised at the foreign governments contracting the American universities for setting up various academic programs and institutions. Ultimately, it amounts to the selling of American thinking, American ways of managing, and organizing systems. American educators in foreign countries are truly the foot soldiers of American Civilization.

Feb 23, 2007/Mar, 2010

COMMENTS

Very good reading. Enjoyed it. **Harbans**

Dear Professor Bhatnagar, I am sorry I cannot send you the needed reference. However, I am forwarding your mail to Dr. Man Mohan who might help. Also, I read all your reflections and enjoy them as they are quite informative and interesting too without exception. For example, even the present reflection contains some information which many people, including me, might not know. At the same time, it is also known that Young did a lot of mathematics as a brilliant mathematician. They say that even germs of Lebesgue measure and integration are found in his work. Of course, he did quite a lot on Fourier series.

Your article may perhaps be very good for a journal on history, I do not know, but about **Ganita Bharati**, I doubt simply because it talks about a well-known mathematician without incorporating a word of mathematics that he did. You will be receiving **Ganita Bharati** Vol. 28 soon as it is being dispatched to you. My most sincere and honest request to you would be to please try to have an idea of what the journal aims at. An article has to be on history of mathematics, and as such mathematics first and history later. It has to have mathematical contents in the historical perspective in an innovative fashion.

There are many occasions when I feel strongly tempted to revert after reading your reflection, but because of my bad habit of putting it off, the ultimate result is empty. Sometimes a draft of an unfinished reply is to be dropped when it continues to be uncompleted for long. Please do continue sending me you reflections. Best regards, **B. S. Yadav**

My grandfather earned his M.Sc. in physics from Calcutta University in 1910. Certainly Bihar did better than Punjab, because the Patna Science College was established in 1915, where my grandfather taught from 1915 through 1945. This college started courses leading up to M.Sc. beginning about 1920. Sir JCBose was one of his teachers. It is true that all three "Presidency" colleges established by the Brits were affiliating universities, and are still so. I hold an M.E. (Engineering) from Calcutta

University, though I was then a teaching fellow at Bengal Engineering College in Howrah.

As in any colony, the real aim of the colonizers was not spreading education, and whatever they did was only to help them administer their colonies. But India did much better than other colonies. I read somewhere that Indonesia had just 17 colleges in the whole country in 1947. **Satya**

PERSONAL REMARKS

HISTORY, MATHEMATICS & HERITAGE

History is a unique subject. The makers of history are seldom its regular writers and readers. The paper brings out several dimensions of history of mathematics, in particular. There are pertinent questions like those proving mathematical theorems have only a tunnel view of the history of their research problems. In general, the hard core mathematicians have downplayed the history of mathematics as a subject of graduate studies. In the academe, history of mathematics is neither a part of history departments, nor of mathematics departments. The fundamentals researches in such a discipline are never possible. There are only a couple of universities in the US that have positions earmarked for the history of mathematics. The paper throws light on its whys and hows.

History, in all its facets, is an index of a nation's total development. Whereas, mathematics may be studied independent of politics, humanities and literature, its history is a patch in the quilt that history is. In Indian context, the paper attempts to survey history of mathematics after independence in 1947, before independence under the British, Muslims, and under the Hindus going back to medieval and forgotten past. It eventually poses the most crucial question on the necessary and sufficient conditions for the development and flourishing of mathematics in the smallest academic unit and a nation, at large.

The nursery of every academic discipline sits in the schools and young children. Mathematics Education has to find a respectable place in the colleges and universities. Mathematics Education or History of Mathematics should not be deemed as a place for the frustrated mathematicians. Mathematics is defined by it abstraction even in the fields that are considered applied. Hence mathematics meshes well in a society where the freedom to inquire, question an authority, a principle, and a book is guaranteed. Finally, the paper makes a bold connection of mathematics with Hindu heritage and religion.

Oct 12, 2008

PS: 5/2014

The above text is an abstract of a paper presented at the annual meeting of History of Mathematica and Heritage conference held in Imphal (Manipur), India – during Dec 19- 21, 2008. However, it merits inclusion in its own right – also because for its unique conference theme.

PERSONAL REMARKS

NEW MATERIALS ON HISTORY OF MATHEMATICS

Mathematics is secular in a sense that it is done 'identically' in every land and culture; whereas, arts, music, philosophy, humanities and social studies are greatly influenced by local conditions. However, history of mathematics is a different world. Certain cultures and religions have contributed, and are contributing far more than others. Therefore, it is pertinent to explore the whys and hows of these questions. For the last couple of years, a focus of my researches has been on the impact of Hinduism on mathematics, and mathematical contributions of the Hindus in India and abroad.

During Dec 19-21, 2008, a national conference for History of Mathematics was held in Imphal, Manipur (India). In an informal gathering, a senior professor bemoaned at the lack of material on Indian History of Mathematics. Generally, research material is interpreted in the form of books and manuscripts. Modern mass printing is only 500 years old; therefore, any search for such books is out of question. Since the 11th century, science and mathematics eclipsed in India, as it was brutally run over by foreign invaders and rulers of different faiths and cultures.

Let it be understood that for the last 400 years, history of world mathematics has been dominated by West European countries. During the last 20 years only, with the growing research in ethno-mathematics, a body of work has come out on the non-European roots of mathematics.

India has a long tradition of using different writing materials, like *bhojpatra*, palm leaves, bamboo stripes, and even barks of special trees. Later on, inscriptions on copper plates were used for official deeds and proclamations. This kind of writing, meant for posterity, was a privilege, and it was financially supported and commissioned by the rulers and aristocrats. Scholars could never afford to buy any material for the sake of their writings, as done today.

Another natural question is that who would have ever paid to have mathematical results written up? On checking with any Indian museum containing manuscripts, or libraries of manuscripts, one would find that they are mostly commentaries on religious and spiritual legends of India. A full day of Dec 23 was spent in Guwahati State Museum, one of the best state museums in the country. All the manuscripts were on Hindu epics of **Ramayana**, **Mahabharata**, **Srimad Bhagwatam**, ceremonial rites, and **Kaali Puja**. These subjects continue to draw the Hindu masses and intelligentsia alike since time immemorial.

Let me add a foot note on Indian manuscripts, as saved by Maharaja of Travencore, when he learnt of their systematic distortion and annihilation by the British in India. The Christian missionaries were the spearheads of colonial military forces. For example, the Spanish church literally destroyed all the ancient records of Inca and Maya civilizations in South America. It then became easier for the Spaniards to rule the natives. Partly due to Hindu caste system, old books and manuscripts are still preserved in some families.

Maharaja of Travancore state outbid the missionaries to procure the manuscripts and scrolls. He converted a part of his palace to a public library that I visited in 1986. India is indebted to this rare nationalist prince. Any way, when it comes to science and mathematics, there is little hope from these sources. On the top, a few scholars knowing ancient languages of the manuscripts, like Pali, Kharosti, and versions of Sanskrit, are miles away from their knowledge of science and mathematics.

The new mindsets have to come forward to decipher sciences and mathematics from magnificent ancient Hindu temples that alone have survived deep in south of India. **Yes, ancient monuments are the books of knowledge. In fact, they are archaeological DNAs**. In a monument, the measurements of various lengths, angles, directions of sun, and other planets can shed light on many aspects of ancient society.

After having taken four archaeological courses in Mexico (2005), Peru (2007), Bolivia (2008), Guatemala and Honduras (2012) organized by the Mayan Exploration Center, Austin, housed in the University Texas, I am amazed how entirely obliterated civilizations are being re-constructed. In India, it would require teams of commercial entrepreneurs and financial investors besides linguists, statisticians and anthropologists. Hindu minds, driven individually, shall face a team challenge first. These complex problems are highly interdisciplinary in nature.

Besides, the study of full scale monuments, the ancient potteries, textiles, coins, clay seals, and terracottas are all sources of new materials for history of mathematics and science. They are well-posed problems, and remind me of inverse problems in mathematics, like that of wave propagation. On the surface, 'crude' mathematics may be in geometrical patterns, but the sophisticated math lies hidden in the chemical composition, knowledge of physical properties needed to develop these materials. It is ultimately tied with the existence of necessary and sufficient conditions for the development of science and mathematics in a society. A good guess can be arrived by methods of extrapolation.

Jan 06, 2009 (India)/July 2014

NUMBER-LESS-NESS OF NUMBERS!

About the natural numbers, so much has been written up and spoken about in every land and culture - in depth, length and width - from their divine creation to logical constructivism. Pythagoras and his school of thought believed that natural numbers unlock the secrets of the universe. This belief is shared by common man everywhere. However, the intellectuals continue to debate over them from different angles. The four *Vedas*, the ancient Hindu scriptures, have patterns of numbers in several *mantras*. For historical lack of interdisciplinary scholarship in Vedic numerology, Vedic Sanskrit, and mathematics, their meanings are not fully unraveled and understood.

This train of thoughts halted at a mental station when, in a casual conversation, my friend, Raju Abraham, an English Professor, told me last week that his birthday falls on 13th. He reeled down a list of good events in his life associated with number 13, or its derivatives – like $26 = 2 \times 13$ and 31 digits of 13 reversed. He excitedly recalled his first job interview taken place on the 13th of a month and in room numbered 13 – happened 40 years ago! In many cultures, Number 13 is considered unlucky, or even neutral. Individual faith of person is a limit of a string of similar events. Of course, it is not the same as a mathematical limit.

Within minutes of this conversation, I received the following e-mail from Cambodia:

"Satish, I'm here in Angkor again. Remember when I asked why 108? I think I have figured it out, but I must be crazy. Look at this and tell me I'm wrong. The Hindu Yuga cycle has four increments:

1 – 1200 years 2 - 2400 years 3 – 3600 years 4 – 4800 years. So…
$1200 \div 108 = 11.11111111111…..$ $2400 \div 108 = 22.22222222222….$
$3600 \div 108 = 33.33333333333….$ $4800 \div 108 = 44.44444444444….$

The repeating decimal denotes the never-ending nature of time and the numbers for each list the four ages, 1 through 4. Could it be this simple? Has anyone else ever mentioned this mathematical link between the Yuga cycle and 108? Baffled, Ed".

[Dr. Ed Barnhart, Maya Exploration Center (MEC), 3267 Bee Caves Rd. Suite 107-161, Austin, Texas 78746. (512) 350-3321"]

Dr. Ed Barnhart and his associate, Dr. Chris Powell, in the MEC, are unique archaeologists who have been involved in researches in the areas ethno-mathematics, geometry and astronomy in non-traditional approaches. They take all possible measurements, whenever, they visit ancient monuments. After all, the books of science and math do not survive a thousand year. Besides, who would ever save them? But monuments - like the pyramids, can last 5000 years. Ancient structures are books in stone waiting to be deciphered by new breed of experts. Through MEC study courses, I have visited Mexico, Peru, Bolivia, and at my urging to Cambodia last year. The next site will be Ujjain, the astronomical and spiritual center of India since ages.

Though these fractions are all related, yet, Ed's observations can be extended further:

$6000/108 = 55.555555555.....$ $7200/108 = 66.666666....$
$8400/108 = 77.777777777...$ $9600/108 = 88.88888888.......$
Finally, $10800/108 = 100 = 99.99999............!!$

Once or twice a year, a forwarded mail hits my inbox on beautiful and almost surreal patterns of natural numbers. Online searches generate plethora of information on 108 –with its mythological and religious meanings, and its significance. They would appeal and satisfy persons of different bents of mind.

In a nutshell, the Number 108 has a very auspicious place in Hindu mythology and life style. But I don't know its real why. The number of beads in a Hindu rosary is 108. In Hindi language, 'Shri' is used before the name of a respectable man. Particularly, a man of spiritual attainments is referred as 'Shri .....108' means Shri repeated 108 times.

Coming back to one's lucky number, its contemplation is no different from writing an essay, say, on lion. One's perception of 'lion' changes with age - from that of a creature to that of a metaphorical one. At least, no single number is lucky for every human being. Does there exist a lucky number for a particular person? Its answer is non-trivial, and makes sense due to the uniqueness of each human being. In almost every country, popular magazines and newspapers thrive on numerological columns. Vedic Numerology is an ancient treatise on this subject and related topics like cryptography. Incidentally, research is always driven by funds - private or state. Mathematicians in academe normally pay little attention to questions rooted in religions.

Personally, I have pondered, played and analyzed natural numbers at different phases of my life. Before studying Number Theory, during 1960-61 for my master's from Panjab University, Chandigarh, I was intrigued by the role of natural numbers in astrology - a corollary of my interest in palmistry. For a number of years, like in many lands, I believed in mystical properties of Number 7. The slot machines of Las Vegas casinos have varied strings of 7's for pay-offs to the winners.

Continuing this thought on Number 7, once, I inquired from HR Gupta (1906-1988), my instructor in two graduate courses and a well-known number theorist, as to whether or not there are any special mathematical properties of 7. His quick and emphatic answer was: "None!" Well, 7 is the largest single-digit prime (prime is a natural number greater than 1 having only two divisors - 1 and the number itself). So what?! To me, it captures the eternal state of natural numbers - above all, the pervasive presence of mathematical forms.

Aug 21, 2011

COMMENTS

As always, very enjoyable--one of my favorite courses in my undergrad was in number theory from Dr. Jim Cangelosi (most famous for his numerous textbooks in mathematics education)... When are you teaching courses this semester? Cheers, **Aaron**

Interesting reading! I think a lot of research has yet to be done in this intriguing area... **Abraham**

Trying to follow this, literally made me dizzy. How are you able to retain these multi-level thought patterns, but so often fumble your routine, single-step, and mundane tasks?? **Gori**

HISTORY OF MATHEMATICS IN PUNJAB
(1849-1959)

Medieval History of Punjab

History of Mathematics in present Punjab is shrouded in the most tumultuous history of Punjab itself. The word Punjab has Persian/Arabian roots - *'punj'* means five and *'aab'* means water. Thus, Punjab means the land of five rivers - Sutlej, Beas, Ravi, Chenab and Jhelum - listed from East to West. They all flow out of the Himalayan glaciers and converge together before submerging into the Arabian Gulf.

The name Punjab tells that its pre-Islamic Hindi/Sanskrit name, *Sampt Sindhu*, is completely erased from public memories. *Sapt Sindu* means the land of seven rivers - including westernmost Sindh River. The 7th river Saraswati, in the easternmost region, has dried out. Furthermore, the ancient civilizations of Harappa and Mohenjo-Daro had flourished in this region. The region had the famous Takshila University. During the last millennium, this fertile and prosperous land has attracted adventurers, explorers, marauders, invaders and traders from Central Asia, Middle East and Europe. For a perspective, India, known as the Golden Sparrow, through the 10th century, was a magnet for the world, as the USA has been for the last 50 years.

In a mathematical jargon, **the necessary and sufficient conditions for intellectual traditions in science and mathematics are peace and prosperity**. This region has not seen stability since Alexander defeated a great Hindu emperor, Pourash (Porus is its Greek distortion) militarily and diplomatically, in 326 BC. In Afghanistan, Islam had over-run Buddhism and Hinduism by the 9th century, and its followers looked eastwards over Khyber Pass for further expansion. India was attacked by different Middle Eastern and Central Asian invaders hundreds of times. Subsequently, it made India porous and weak.

The foreigners annihilated and plundered the Hindu institutions and temples – carried out untold massacres and conversions of the Hindu populace. Thus, during the 14th to 18th centuries, the region witnessed only great generals, soldiers, armaments and battles. The four major battles fought in the plains of Panipat alone. This political unrest scattered away the Hindu masses and institutions into wilderness in their homeland.

Modern History of Punjab

The first historical turnaround in Punjab came with the founding of the first Sikh Empire by Maharaja Ranjit Singh (1789-1839) when the waning Muslim rulers of the region were thoroughly defeated. However, the vast Sikh Empire politically collapsed soon after his death -mainly for the lack of political vision. The British vanquished his successors and annexed Punjab into the British Raj in 1849.

The present Indian state of Punjab was created in 1966 out of the Punjab state which itself was created after independence in 1947. It then included the present states of Himachal Pradesh and Haryana. However, at the time of partition of India, Punjab was divided - nearly two thirds went to Pakistan along with its capital city, Lahore, and the home of Punjab University (PU). PU was established by the British in 1882 as the 4th university in British India. However, Mohindra College, Patiala, in Punjab, was established in 1875. Its academic affiliation was then changed from Calcutta University to Punjab University. The College was a gift to the Maharaja of Patiala for supporting the British against the nationalists in the first militaristic freedom movement, also known as the 1857 Mutiny in the British annals.

It is pertinent to add that the school and college curricula in government institutions followed a system of education conceived and implemented by Lord Macaulay (1800-59) for India. It only encouraged British history, British economics, and British philosophy, and British psychology – besides, of course, English literature. Mathematics was encouraged, but sciences seldom. The reason being science is very expensive; secondly, the British wanted the Indians to continue living in their superstitions. It is easy to govern people, who are divided and ignorant.

Mathematics Enterprise in Punjab

As a manageable research project, it is better to focus on the world of mathematics in Punjab, during 1849 - 1959. I am certain that early mathematics lecturers/professors who taught in Mohindra College, Patiala and in PU/Lahore must have been British. Reason - they used to undertake long sea travels to India for hefty salaries and benefits including hardship allowances. Indians were paid less than one-tenth of the salaries of the British in India! Of course, the salary differential applied to their services in any military or civilian jobs. Freedom is most precious in the political life of any society.

I would like to spend a week in PU Lahore and dig into old mathematical records from its archives. It is challenging and interesting problem in the History of Mathematics in Punjab. Here is a math history trivia. As the assets of Punjab University, Lahore were being divided in 1947, in naming corresponding new university of India, the letter 'u' was replaced with 'a'. That is how Punjab became Panjab University, Chandigarh. Since a University Senate /Syndicate has no jurisdiction over the names of the state, Punjab state is spelled the same in both the countries!

It must be added that the Camp College (CC), Delhi, started in 1947, was affiliated with PU which was administratively located in Solan. The CC played an historic role in turning out fine mathematicians. More students finished their MAs from CC, through the 1950s, than turned out by the rest of the PU sites; namely, Ludhiana, Jalandhar, Patiala, and Hoshiarpur - all put together. Mathematics Dept shifted from Hoshiarpur to Chandigarh in 1957. I don't know when the CC was closed, or taken over by Delhi University. It is an open question for research in history of mathematics.

Affinity between Math and Hindu Religion

There is a very unique aspect of Punjabi men and women in mathematics. Statistically speaking, all top mathematicians are Hindus, despite the fact that in north India, the Hindus had lost their freedom for a thousand years. Their collective will being broken, in order to survive political persecution, either they ran into the villages or stayed in subservient roles in the cities – paying double or triple in taxation. The Hindu intellectuals served in the Muslim courts as astronomers, surveyors, traders, and builders. They continued to offer these services to their British rulers, who gave them the titles of Rai Sahib or Rai Bahadur etc. for their loyalty. My great grandfathers had such hereditary titles. I was proud of them for many years till this title piece of history and politics opened my eyes.

The freedom of thought that the Hindu religion encourages in individuals in questioning any hypothesis, belief, book, scholar, or scripture is carried over to fundamental groundbreaking abstract concepts in mathematical researches. To a large extent, its converse is also true. It is reflected in the number of Hindu students going for MAs and PhDs in mathematics – just for the love of knowledge. Moreover, there is an historic connection between mathematics and theology, as supported by the top European mathematicians of the 17th and 18th century who were great theologians too - including Sir Isaac Newton!

Math Professors, Authors, Researchers and Administers

What is history without any names? The following are a few names in the order they came up to my notice. Of course, the list is a work in progress.

1. Hem Raj and Hukum Chand (early 20h century): mentioned in one breath; well-known lecturers in Dayal Singh College, Lahore. They co-authored a classic undergraduate textbook which I used during 1957-59.
2. Hans R. Gupta (1902-1988) MA, PhD/PU: number theorist, Weirstrass of Punjab, a great and compassionate teacher, administrator. Taught geometry in MA Part I in a very unique manner; impacted me professionally and personally. IMS President
3. RD Syal: taught Astronomy in the Ludhiana site of PU through the 1950s. (Relativity was also one of the papers in MA. I studied it independently after MA.)
4. Sarvadaman Chowla (1907-1995) MA/ PU, PhD (Cambridge) under Littlewoods: Well-known number theorist. His father, Gopal Chowla was also a math professor with PhD (Cambridge);

and worked in PU Lahore. I met him in Laramie in 1984 through a common acquaintance.

5. SD. Chopra (1915 – 1987): Great teacher, researcher and administrator; built a strong Math dept at Kurukshetra University (KU), IMS President. During 1965-67, I worked under his guidance for PhD (unfinished) in Mathematical Seismology.

6. Mohinder S. Cheema: MA (1950)/ PU, PhD/1961 US: number theorist. Met him twice in Tucson, Arizona.

7. Ram Parkash Bambah (1925-) MA/ PU, PhD/Cambridge, number theorist, top administrator, vice chancellor PU, Padam Vibhushan. IMS President, INAS, socially active. Taught me number theory during 1960-61.

8. Madan L. Puri (1929 -) MA/CC, 1948; PhD Berkeley, well-known statistician, His younger brother did PhD in applied Math from Courant Institute. (A good friend for 40+ years)

9. Bansi Lal: DAV College Jalandhar, a pioneer in writing undergraduate textbooks in math.

10. Shanti Narayan: a self-taught mathematician; the first Indian author to write excellent graduate math textbooks on Modern Algebra, Analysis, and Topology etc. Served as principal of Hindu College, Delhi and was an Arya Samaj leader. I met him a couple of times with Swami Deekshanand Saraswati, my maternal uncle.

11. ML Gogna (1936-2010) MA/CC, PhD/ Cambridge: professor at KU. A great friend, I have ever known!

12. IBS Passi (1939-) MA/PU, PhD/UK: Algebraist, IMS President, INAS.

13. Satish K. Aggarwal (1939 – Present) MA/PU, PhD/US: Professor/MDU.

14. Sarvajit Singh (1939 -) MA/Agra, PhD/KU: researcher, administrator, IMS President, INAS.

15. Bhushan L. Wadhwa (1939 – 2004) MA/CC, PhD/US: Professor and Chair, socially active.

16. Manmohan Singh Arora (1937 -) MA/CC, PhD/US: Professor

17. Jitendra N. Manocha (1937 -) MA/CC, PhD/US: Professor. His late wife, Dinesh also did MA from CC and PhD in Operation Research from the US.

18. Kehar Sigh (1936 -) MA/PU, PhD/KU, Professor/GND.

19. Satish C. Bhatnagar (1939 -) MA/PU, PhD/US: Professor at UNLV, Reflective writer; History of Math, Math Education.
20. Daljit S. Ahluwalia (1933 -) MA/PU, PhD/US: Researcher, NJIT; known Chairman.
21. DS Gill (1940 -) MA/PU, PhD/ US. Professor/Cal State, Pomona.
22. Dalip S. Saund (1899-1973) MA/PU, PhD/US, first Asian to serve in the US House of Rep.
23. Ram Tirath (1873-1906) MA/PU. Inspired by Vivekananda, preached Vedanta; quit on math.
24. Manohar L. Madan (1931-2011) MA/CC, PhD/Germany, Researcher/Algebraist.
25. Ranbir Singh Sidhu (1940 -) MA/PhD from KU, was my student during my first year (1961-62) of college teaching in Bhatinda. Retired from GND

Curriculum History

For a bachelor's degree, Math A and B Courses formed a unique math curriculum in Punjab, as it did not exist anywhere else in India. It was introduced when PU Lahore was established in 1882, but was phased out by the 1960s. Its disadvantage was that there was no breadth in the degree program, as besides English and math, nothing else was required. Nevertheless, it gave a lot of depth for a few lucky ones who went overseas for PhDs. Also, Honors in Math was awarded after passing three additional math papers - besides four regular three- hour exams in Math A and B courses. It was unique too - being different from Distinction in Mathematics - common in most Indian universities. A history of this math curriculum is an open question – how and where it was drawn from?

To some extent, this math curriculum was a watered down version of a math program of Cambridge University. It was 50 % math and 50 % physics – to be called mathematical physics by today's US standards. The US colleges don't have mathematical physics as a popular program at BA or MA level. Interestingly enough, Ramsey's textbook on Statics and Dynamics is still seen being used in some universities in India - after over a century, a relic of academic bondage!

Mathematical Snippets

During the 1950s, the state of mathematics culture in Punjab can be envisioned from my experiences. When I did my BA (Hons. in Math) in 1959, my two college instructors had no idea of what PhD in math meant! They had vaguely heard that in PU Chandigarh, some lecturers had PhDs. After joining PU Chandigarh (from Bathinda via Delhi!) for my MA in Math, eventually, I noticed that most dept heads and faculty members were from the state of Tamil Nadu. It was due to AC Joshi, the second Vice – Chancellor of PU, who recruited talent from anywhere.

The 1947 partition had created unprecedented social and political turmoil due to a movement of 10 million people across the India-Pakistan border. Their settlement took all national resources. Punjab had no serious college education before the 1960s. Research had absolutely no meaning to the general public. My mother, bless her soul, often wondered at my studying after MA! The academic parochialism raised its ugly head only in the 1960s, when the states of Haryana and Himachal Pradesh were carved out of Punjab. Subsequently, some natives of the states earned first PhDs. The parochialism was also compounded by scarcity of jobs at that time in India.

Mathematics, in Punjab, still has a long way to go. Generally, in a poor or developing country, the youth follows a career path where money is flashing. Personally, I was all set for the elite Indian Administrative Services, but a quirk incident changed the course of my life. I joined the ranks of college teaching in 1961, in my hometown Bathinda. Yes, it has been great – 50+ years!

As industry and computer technology are sweeping the world, the quality and quantity of students going for math degree is falling. Surprisingly, in all institutions that I have visited in India, the number of women doing graduate work in math is at least twice that of men. **How do you explain this gender incongruence in math?** Nevertheless, future of math does not lie in its standing by itself, but in making connections with sciences, technology, arts and social studies.

Conclusion

In 1882, the entire India had only four universities and perhaps a dozen colleges. Today, there are 30 universities (not as comprehensive as in the US) and nearly 300 colleges in Punjab alone. What an explosion in educational institutions both in public and private sectors! Education - from pre-schooling to tertiary, is a driving engine of the economy of India, the only country in the world.

Some Open History Questions: However, when it comes to a day when a Punjabi, living in Punjab, ever winning a Fields Medal or Abel Prize, the highest mathematics research honors in the world, it has to wait for another 100 years. **Reason**: Deep scars of poverty and subjugation of centuries have to be erased from the heads, hearts and minds of the Hindus who constitute 80% of India's population.

In its data mining aspect of this project, I am keen to know the bios of the vice chancellors and heads of math dept of PU. My conjecture is that they were mostly English before the 1920s. Any native has to be the English lapdog.

A week ago, I approached a lecturer in Pakistani Punjab who got earned MA from its PU to get involved in having a similar data from that region. Also, I will circulate it to the PU Alumni, as some one may pitch in some historical tidbits

Sep 24, 2011/July, 2014

PS: This ongoing ***Reflective project*** is an offshoot of an e-mail received in Sept, 2011 from an old-timer friend, IBS Passi. He was then invited to deliver a talk on this subject at the Ramanujan Mathematical Society's annual conference held in Allahabad, India, during Oct 2-5, 2011.

COMMENTS

1. Dear Satish: Thank you for sending me highly interesting article on Punjab Math. Best regards, **Madan Puri**

2. Very well written—**Sarvajit**

3. Dear Prof. Bhatnagar, I have yet to read your write-up carefully, but I noted some immediate points-I don't know when the CC was closed or taken over by Delhi University. You can ask Prof. Man Mohan Sharma on phone-he is at Atlanta-Vice-President, Clark Atlanta University (If you do not find his phone numbers, you can call his Delhi office-91-9811102512. He (and Prof. M.L. Gogna completed their M.A./M.Sc. together in 1962, I think and then it was under University of Delhi Post-graduate Evening Dept./ College www.du.ac.in (University website may have details too). Maybe after talking to him and seeing his CV, you would like to include his name? I cannot figure out your benchmark for the list of names as it stands.

   Ram Parkash Bambah (1925-) MA/ PU, PhD/Cambridge, number theorist, top administrator, Padam Vibhushan. Taught me number theory. IMS President; socially active. He has many more important academic and administrative achievements. Maybe, you can look for them on the web or ask him or Prof Passi. (Even Prof Passi's portion needs to be enhanced).

   Bansi Lal of DAV College Jullundur, a Punjabi pioneer in writing undergraduate textbooks. Do you mean Prof. Bansi Lal, who taught at the Kirori Mal College for decades together? He lives in Delhi (1, Vaishali, Pitampura, New Delhi (can see the code on web) and is still active.

   Daljit S. Ahluwalia (1933 -) MA/PU, PhD/US, Active Researcher and able Administrator. For uniformity, you can tell his Institution-NJIT (full name, in fact).

   It seems this write-up will undergo some evolutions. Best regards, **Ajit**

4. Dear Satish, as usual good work! Have you ever wondered why Indians planted in the Western soil grow into stalwarts while back home they get stunted? **Harbans**

5. Dear Professor Bhatnagar, Your account of a century of Math in Punjab is extremely well written and perceptive. It has personal appeal for me as I now know something about my friends who were working for Ph.D. in PU when I was there in 1961-62. **ARUN VAIDYA**, AHMEDABAD

PERSONAL REMARKS

CULTURE - CONFERENCE COMBINE

In contrast with Indians, the academic conventions, conferences, meetings and seminars are primarily the products of the organized western mind. A few academic societies and associations were formed in early 20th century of pre-independence India to transplant limited British intellectual traditions. After decolonization, particularly since the 1990s, there has been a sizzling growth in the number of private and government colleges and universities in India. Proportionately, the professional organizations have neither influenced their disciplines nor expanded their activities. For instance, Indian Mathematical Society (IMS) was formed in 1907, but its annual meetings hardly draw more than 100 participants – twice, I have witnessed its 'show'. Actually, more Indians go overseas for the US math meetings!

The state of Indian mathematics organizations is a reflection of the present individualistic mindset of the Hindus who constitute 80% of India's population. Generally, the societies are founded by individuals – thus, they remain personality driven, rather than steered by broad missions. Besides, the IMS, India has a dozen of other math organizations. However, they are all anemic, as they look relatively bonsai in their growth.

The other main reason is that Indians think that research and scholarship in mathematics are one man activities. For years, I too subscribed to this philosophy. Only recently, it dawned on me that science and mathematics are super-organized disciplines, as a result, collaborative research, whether tight or loose, truly promote fundamental researches. Conferences provide a stimulating environment. Again, this individualistic thinking comes from the present Hindu religious beliefs of having individual little temples in homes, personal deities, gurus, and what not! It is time to cultivate collective training of body and intellect as early as possible.

During Jan 09-11, 2012, I attended the annual meeting of the International Society for History of Mathematics (ISHM) in V V Nagar, Gujarat, India. It was hosted by one of the few engineering institutions in

India - opened exclusively for women. Nearly 100 persons, including students, participated. Only three participants came from overseas – including me from the US, one from Canada and the third one from the UK – all originally from India. For me, this visit served three purposes – professional development, strengthening ties with friends and relatives, and thirdly drawing spiritual nourishment from its ancient heritage. Incidentally, it was my third participation in the ISHM meetings - the previous ones were in Indore (2004) and Manipur (2008).

A few cultural features typically factor in the organization of Indian conferences. There are pompous opening and closing ceremonious of the conference taking two hours each – but having little to do with anything mathematical. Apart from institutional dignitaries, community leaders are also invited. I was one of the thirty invited speakers. All were ceremoniously recognized on different occasions more than once, and presented gifts, honoraria and certificates accordingly.

The research papers covered the entire spectrum of topics –related or unrelated with the theme of the seminar. One idea behind the conference site is to give national exposure to the under-represented regions of India. Likewise, the paper sessions did not adhere to the time slots – they were easily out of phase by 20 minutes in a 2-hour session! It may be attributed to 'Indian time' management. Philosophically, time is perceived differently in Indian subcontinent. The speakers keep on speaking irrespective of the mood of the audience. I am indebted to the toastmasters for having learnt time management in public speaking.

There being no designated area for informal meetings and exchange of ideas, it was funny to see spouses and visitors sitting in front rows - without having any knowledge of mathematics. In contrast, it did remind me of US math colloquia and paper sessions attended by only a couple of participants, who actually go for the specific topics.

Interestingly, what sets an ISHM conference apart from a US math meeting is that the invited guests were housed free, served bed tea in their rooms - followed by breakfast, lunch and dinner at the conference venue! It was so heartening to see undergrad student volunteers making sure that the services were good and prompt – unthinkable in the present US culture! They personally helped me in locating rooms, securing my travel bags, preparing a slide of a document, and finding a place for internet usage. Informally, I spoke with a group of students and distributed cards and bookmarkers of my book, *Scattered Matherticles: Mathematical Reflections, Volume I* and that of UNLV Math Dept that I had carried for recruitment and publicity for its MS/PhD programs.

Furthermore, transportation was available from the university guesthouses to the conference site. On the second day, there was one-hour cultural program performed by the engineering students. It was impressive to see how some students were so talented in dance and music. The cultural program was held on the premises- right after dinner. A local sightseeing tour – including the world famous Amul Diary in nearby Anand, was scheduled on the third day, but I had to leave just before it was to start.

On a personal note, I renewed some professional acquaintances, established new contacts, and witnessed an incredible growth of higher education in Gujarat, the fastest developing state of India. My talk was on the lines of David Hilbert's historic talk given in Paris at the 1900-ICM Meeting (International Congress of Mathematics), wherein he gave out 23 unsolved problems in hard-core mathematics. According to Goggle search, ten problems are resolved, three 'unresolved' and ten are partially resolved.

Essentially, I talked about my lifelong project of 52 + 4 problems on the History of Hindu Mathematicians, and also presented my **Ace** solutions to four of them. There is absolutely no finality of solutions outside the world of hardcore mathematics. Overall, it was a satisfying round trip from Las Vegas to VV Nagar –considering over 20,000 miles and 80 hours it took!

Jan 16, 2012/Aug, 2014

COMMENTS

Thanks. Quite interesting comparisons. **Raja** (Retd Physics Professor in Chicago)

Dear Professor Bhatnagar, Thank you for the very interesting, extremely perceptive and extra ordinary report on the ISHM Conference. You have hit the nail on the head in many of your observations. Hope to receive many more such pithy and to-the point writings. **Arun (Vaidya).**

Dear Prof. Bhatnagar, It is really pleasant to read your views. Conferences here are like 'Mushaira' of poets, where no speaker (Shair) listens to others and recites 'Wah-Wah' as a holy tradition.

Another development is the condition (from UGC, India) to score some essential research points for promotion of college/university teachers to higher scale/slab/position. All types of participation in these conferences yield some contribution. Hence, we can see many get-togethers to be declared as conferences (obviously National conference). In the mail you have mentioned 52 + 2 problems on the History of Hindu Mathematicians. I wonder if I could get a glimpse of few such problems. Thanks for everything. Kind regards from **MD Sharma** (Kurukshetra U)

Dear Professor Bhatnagar, Thanks for your thought provoking message highlighting your views on the recent ISHM conference. I am also thankful for your sending me regular interesting attachments on Mathematical Reflections. We at Ramjas College, Delhi, are organizing an International Seminar on History of Mathematics on Nov. 19-20, 2012. The delegates will then collectively go to participate in the ISHM Intl Conference at Rohtak. I take this opportunity to invite you to participate at Ramjas Seminar. Ramjas College has been the breeding ground for ISHM ever since it was founded. Regards,
Man Mohan (DU)

Interesting Satish! Am not sure whether the unsoloved Hilbert problems are the numbers from the Google. **George**

Hi Satish, Your note is an in an interesting read. This should useful for future event-organizers. I hope you have copied to the lady from Rohtak, where next ISHM is being planned. I have similar opinion of beginning and "end" of the professional societies. My guide, late Prof. Pramila Srivastava (a student of B N Prasad) put my name in some capacity for Allahabad Math. Soc. for 'decorative' purpose. I have never attended any meetings or events. Cheers. **Hans Joshi**

Dear Satish Ji: Happy to read this account. If you have paper presented in Gujarat, please send me its copy. **BhuDev Sharma** (well-known intellectual)

Dear Prof. Bhatnagar, I have cc-ed your e-mail to the IIT (B) faculty. This year there were about 400 participants for the annual IMS conference, and about 150 for the annual RMS conference. The fact that there are three major math societies in India, which "compete" with each other for govt-level honors. No society actually represents the Math community for their memberships are not significant. You may have better perspective on this which you wish to share with me and others like Profs Dani, M. S. Raghunathan, M. S. Narasimhan, C. S. Seshadri and others. Warm regards, **Ravi Kulkarni**

DECKING OUT HINDU MATHEMATICIANS

INTRODUCTION

This is a long-term collaborative project on the ancient history of mathematics which was conceived five years ago when I was attending the JMM-2010 at San Francisco. While searching for great mathematical minds of the yore, the project also attempts to reconstruct the history of the rise and fall of Hindu centers of learning where seminal works in science, mathematics and astronomy were undertaken. For instance, the three great ancient universities, *Takshila* (now in Pakistan) *Vikramshila* and *Nalanda* (both in the present state of Bihar, India) flourished and survived for a thousand years before they were destroyed and burnt down by the armies of the Muslim invaders, namely by Mohammed bin Bakhatiyar Khilji, a general of Mohammed Gori of Afghanistan.

These universities were renowned, and comparable with the present-day US universities – like, Stanford, Harvard and Berkeley. Besides, there were hundreds of smaller centers and academic institutions which are also lost and lying buried. They can be traced and tied with the golden period of the Hindu kingdoms which lasted from the 6th to 10th century AD.

The decline of Hindu culture and civilization started with internal social divisions and political dissentions. India is a perfect example of a nation that first collapsed from within. The outside powers who recognized it took full advantage of internal discords. Thus started hundreds of attacks by marauders, invaders and rulers from Middle East, Central Asia and Europe. This made India eventually very porous in all fronts. Inevitably, it resulted in a long period of subjugation of the Hindus - in their homeland, India.

There was a total annihilation of Hindu institutions, places of worship, and suppression of their faith and beliefs. It has continued through the middle of the 20th century. As a consequence, during the last one millennium, the pursuits in mathematics and sciences have been replaced by mindless superstitions and blind practices – going rampant even after independence in 1947.

VISION

The goal of the project is to resurrect a total of 54 mathematicians - 13 mathematicians from each geographical region of India – east, west, north and south - from the ancient times to the end of the 12th century. The additional two are open to all four regions. The defeat of Prithviraj Chauhan (1149-1192), the ruler of two vast regions around Ajmer and Delhi, was when the sun began to set on the Hindu kingdoms of India. That is a rationale for the time line ending at the 12th century.

The project also aims to touch upon the impact of major religions and political systems in the development of mathematics and related areas in India. It is at once labor intensive and intellectually demanding. Preliminary reports have been presented in mathematics meetings and ideas discussed in a history of mathematics course (MAT 714). The ultimate vision is to ignite a flame of intellectual renaissance amongst the Hindus – through a mathematical window or paradigm.

Some of the following material is adapted from a paper presented at the annual meeting of the International Society of History of Mathematics held in New VV Nagar (Gujarat), India, on Jan 9-11, 2012. I opened my talk in the spirit of the presidential address that German mathematician, David Hilbert (1862-1943) gave at the International Congress of Mathematicians held in Paris, in 1900. Hilbert put forth a list of 23 unsolved problems - broadly in the areas of analysis, geometry and foundations. They are reckoned the most successful and deeply considered compilation of open problems ever to be produced by an individual mathematician.

However, I added that Hilbert brought out his 23 unsolved problems at the age of 38 and I am putting forward 54 open problems in the history of mathematics of ancient India at the age of 72. It must be added that in mathematics, by and large, an open problem can be stated sharply enough – to be proved right or wrong. In the history of mathematics, generally there are disagreements about a fact being right or wrong.

DECK OF MATHEMATICIANS

Where does the Number 54 come from?

The number is taken from an ordinary deck of playing cards that has 13 cards each in its four suits, plus 2 non-playing extra cards, generally called, jokers.

WHAT DOES 4 CORRESPOND TO?

It correspond to four geographical regions of ancient India – namely, south, east, north and west. In order to give a Las Vegas touch to the project, the Spades, Clubs, Diamonds, and Hearts, used in a deck of cards, are aligned with the south, east, north and west of India respectively.

WHAT DOES 13 CORRESPOND TO?

Mainly, the project is about resurrecting 13 top mathematicians in each geographical region – their personal lives, social and political conditions in which they thrived, and mathematics they created and/or applied. Moreover, the mathematicians are to be ordered from the highest in terms of their intellectual contributions The list begins from the Aces of cards for the highest at Number 1, then Two of cards for Number 2, and so on – thus, King for the last, Number 13.

It may appear that this project of finding 13 mathematicians, say, from south India, during a span of 2700-2900 years (1700 BC to 1200 AD), should prove to be an easy undertaking. After all, in such a vast ocean, any net would catch some fish. However, the problem is that there are literally huge black holes in the history of ancient India, as it was systematically decimated by foreign rulers, who believed that it would be easier to rule over the Hindus, if they were disconnected from their glorious past. Simultaneously, they forced their foreign culture upon them. That is why one finds only scarce and scattered pieces of historical records covering this vast period.

That is precisely a point where archaeology, politics, social traditions and religions kick in. Mathematics and sciences have to be extracted no less from ancient monuments whose remnants are still standing – apart from their applications in astrology, astronomy, *Jyotish* (prediction theory).

THE ACE MATHEMATICIANS

During the talk, I dramatically declared the solutions of 'first' four out of 54 problems. They are the following four Vedas, the most ancient and sacred scriptures of the Hindus:

1. *Rig* Veda represents the South of India, and corresponds to the Ace of Spades. Its 'authorship' or divine revelation is attributed to Agani Rishi, Agani as a concept, or Agani as a school of thought. Rig Veda has 10,552 mantras and contains 21 branches of knowledge.

2. *Atharva* Veda represents the East of India, and corresponds to the Ace of Clubs. Its 'authorship' or divine revelation is attributed to Angira Rishi, Angira as a concept, or Angira as a school of thought. Atharva Veda has 5977 mantras and has 9 branches.

3. *Saam* Veda represents the North of India and corresponds to the Ace of Diamonds. Its 'authorship' or divine revelation is attributed to Aditya Rishi, Aditya as a concept, or Aditya as a school of thought. Saam Veda has 1875 mantras and has 1000 branches.

4. *Yajur* Veda represents the West of India and corresponds to the Ace of Hearts. Its 'authorship' or divine revelation is attributed to Vayu Rishi, Vayu as a concept, or Vayu, as a school of thought. Yajur Veda has 1975 mantras and has 101 branches.

That is about the aces, the topmost four mathematicians from four regions of India. One may raise an objection that they are books, not the human beings. Well, books are the dead persons and persons alive are the living books. Also, one's faith enters here forcefully, but stealthily. For example, the holy scripture of the Sikhs is called *Guru Granth Sahib,* the manner in which a person is respectfully addressed. The Sikhs believe in it, as the 11th eternal Guru, and revere it accordingly. Likewise, the knowledge contained in each Veda is beyond the scope of one individual, and yet revealed through a rishi, an enlightened yogi. It is like the *Quran* as revealed to Muhammad and the *Book of Mormons* to Joseph Smith.

In Christianity, Jesus, claimed as the son of God, is "the Word" ("the word was made flesh and dwelt among us" John 1: vs 14), and the Bible is also called "the word of God."

GIST OF VEDIC CORE

I must say that the idea of Ace mathematicians was a revelation to me, and it flashed like a bolt from the blue. I felt startled. Let me also add that you do not pick up a Veda and start reading it. It is like any 10-year old kid reading a graduate math textbook, would have no idea about mathematical notations and concepts therein. In order to have a degree of mastery of math, a person has to study it in high school, college and university.

Likewise, there is a systematic study of the Vedas. It begins with **six** *Vedangas*, namely:

1. *Shiksha* 2. Grammar 3. *Nirukta* 4. *Chhanda* 5. *Jyotish*, 6. *Kalp* – followed by

Six *Upangs/Darshanshasttra*: 1. Nyaya by Gautam, 2. Vaisheshik by Kannad Muni, 3. Sankhya by Kapil Muni, 4. Yoga by Patanjali, 5. Vedanta by Vyas Mini, 6. Mimansa by Jaimini Muni;

Four *Upvedas*: 1. Ayurveda of Rigved, 2. Dhanurveda of Yajurveda, 3. Gandharveda of Samveda, 4. Arthveda of Atharva Veda.

MATHEMATICAL MANTRAS

All the Vedic mantras are coded with layers of multiple meanings – like, the old 4-track and 8-track tapes that recorded different music on each width. At times, some Vedic mantras have various patterns of numbers. They are very simple, but their survival for thousands of years indicates its unknown or lost deeper meanings. Unfortunately, there are no Indian scholars who have sufficiently grasped Sanskrit and mathematics needed for seminal researches. The reason being that even after India's independence in 1947, to the best of my knowledge, no Indian university curriculum allows the study of Sanskrit along with that of mathematics and sciences – be it at a school, college or university level. What a sorry state of intellectual development!

For the sake of reference, a few Vedic mantras each one having a mathematical meaning are given below: They are taken from the *Sciences in the Vedas* by Vaidyanath Shastri (1970).

Atharva Veda: V.15 1-11, IXX. 43. 35, VIII. 2-21
Rig Veda: I. 84. 13, VIII. 46. 22, X, 52. 6
Yajur Veda: XVII.2, 24, XVIII, 25

It must be added that the *Vedic Mathematics* by Bharat Krishna Tirath (1965) contains 18 mathematical sutras (aphorisms) which he claims to have derived from the Vedic core. Well, that means 4 problems are down and 50 more to go!

PROJECT EXECUTION

Great ideas come into the minds of ordinary individuals too. But not all great ideas get enough traction in order to make a tangible impact. A supposedly lofty idea has to be communicated, debated, translated into actions, marketed, and publicized in a recursive manner. That means a critical mass is needed before fusion. Apart from my own initiatives, last year, I approached the Indian Mathematical Society with my funds for instituting two new awards in the ancient history of mathematics. My proposal is still under consideration. For the American Mathematical Society and Mathematical Association of America, I plan on organizing special sessions during their annual sectional meetings.

The first such session will be held during the April-2015 meeting of the Western Section of the AMS. UNLV will host the meeting and I will be an organizer of the History of Mathematics session. Of course, any support from any individual or institution will be always appreciated.

May26, 2010/Sep, 2014

COMMENTS

That is great. All the best. Just a small comment, mathematicians are not evenly distributed in space and time. **Sarvajit Singh**

My God, Professor Bhatnagar, you are a repository of knowledge and a great visionary. Godspeed to your grand "54" plans. I first doubted if you can accomplish your mission 12000 Kms away from India. But then in your case, all bets are off! Wish you Godspeed. ---- **Arun Vaidya.**

P.S.: On only a slightly lighter side, I wish you would meet our PM who is on a visit to your country. When you talk about your plans, I am reminded of him!

"One may raise an objection that they are books, not human beings. Well, books represent the lives of dead persons, and persons alive are living books."

Comment: In Christianity, Jesus is known as "the Word" ("the word was made flesh and dwelt among us" John 1: vs 14) and the Bible is also called "the word of God." Writings capture the essence of the soul of a person; this being especially so when we consider that it is language which distinguishes man from the rest of the created order. As it is the soul rather than its bodily encasing which constitutes the essence of the human being, a book, far more than any mummified cadaver, serves to preserve an individual's human attributes in the most effective way possible. Are not books and blogs the new "mummies" and are not our libraries and the World Wide Web the new pyramids?! **Francis**

SECTION IV

SMORGASBORD BITES

PHILOSOPHY TURNING MATHEMATICAL

Yesterday, I attended a colloquium given by David Sherry, a philosophy professor from Northern Arizona University. The Title of the talk, *Bayes' Theorem and Reliability,* and a line in the Abstract, *'Numerous psychological experiments appear to show that people are not very good at inductive reasoning,'* pulled me to the lecture. The concepts of induction in mathematics and physics being very clear, I was also curious about its philosophical angle.

Currently, this issue seems to be hot in the law courts and clinical psychology. For math students, the scenarios of word problems are not important while applying mathematical principles to work them out. We let the chips fall and move on to the next word problem. The context of the problem is absolutely irrelevant. No human values or judgments are called upon in solving a mathematical problem, or even in analyzing its solution.

Bayes Probability goes back to Thomas Bayes (1702-1761) a British priest whose father was also a priest. Sherry's lecture was on conditional probability and Bayesian Formula. In a finite mathematics course, this material is covered in 2-3 lectures. Sherry presented one scenario from NBA (National Basketball Association) and the other, a court witness. Then he raised questions of reliability. In the domains of philosophy and psychology, it often becomes a ping pong of ideas. It happens because the heart does not accept mathematical solutions, and the head is not equipped to discover new mathematical tools.

During the lecture, my thoughts were bouncing back and forth. The first observation was that **the western philosophy is no longer what its Greek heritage conveys**! Twenty five years ago, I attended a philosophy colloquium delivered by a vice president of an international organization of philosophy on some topic of 'identification'. He made its far-fetched connection with an integral equation! Since then, the study of philosophy has become highly mathematical.

The colloquium attracted nearly 30 people on a Friday afternoon. One half of the audience was from psychology and the other half from philosophy.

The funny thing is philosophy and psychology departments at UNLV do not require even one serious math course beyond a baby course on the *Fundamentals of Mathematics* (MATH 120) taken to satisfy the University's General Core Requirements. In order to keep up with a growing mathematical usage, at least a course on calculus should be recommended. The understanding of deductive reasoning of mathematics is very pertinent.

Thirty five years ago, a well-known Indiana University (IU) psychology professor gave a colloquium in the IU Mathematics Department. He applied the concepts of set theory and mappings in modeling an area of psychology. I was then a recent graduate student from India; it was a breathtaking experience for me to see such an application of mathematics. In the audience were mathematics legends like Zorn, Hopf and Halmos who had left their marks in respective fields. Since then, Mathematical Psychology has matured into a respectable discipline.

The inroads of science and mathematics, started by the minds like, Descartes and Pascal in the 17th century, have propelled western philosophy into a mathematical orbit. In the US, western philosophy can be described as highly utilitarian. At the other extreme, Hindu philosophy seems divorced from life.

Recently, as a member of Asian Studies Advisory Committee, I met a Michigan University Professor of Buddhist Studies over a group luncheon. There, an English professor remarked, "Is there an Asian Philosophy?" The question appalled me - either she had no idea of geography or history of Asia, or forgot to include a specific country. I simply said that the Hindus had gone to the extreme of philosophizing everything in life!

There is a gulf between what is considered philosophy as a discipline in the US and what it is in India. The questions relating to soul, heaven and earth, mythology and symbology, ethics and morality integral to Hindu Philosophy are inseparable from Hindu religion. They have been out from the domain of the western philosophy since the 18th century.

Another thought was that if I have to pick up **one feature that distinguishes the American Civilization from the rest of civilizations in history, then mathematics says it all in one word.** Today, even serious mathematics is percolating disciplines hitherto were light years away from mathematics. I feel good about my own observation!

Don't get me wrong, I am not for more mathematics. However, I am for the right use of math. Math being binary in nature has more limitations than literary languages like English and Sanskrit have them. **Math is rational in approach while philosophy is supposed to be non-rational.** Mathematization of a problem only provides a very different perspective that necessarily may not be the best.

There is a downside of math visibility. For example, a few years ago, a colleague in the Criminal Justice Dept. showed me two manuscripts of the same paper. One, having no math, was rejected by a journal. However, the second one, re-submitted after including a table of data, was accepted by the same journal! Since then, the 'abuse' of math, and statistics in particular, has gone very far that new courses and disciplines are mushrooming up in academia.

In the US, the pursuit of philosophy is at a bifurcation point – like, economics divided into quantitative and qualitative. There is a general 'theorem': Social Studies + Mathematics = Social Sciences. Amongst mathematicians, there is a growing interest in the philosophy of mathematics. Since 1995, a special interest group in Philosophy of Mathematics, under the umbrella of the Mathematical Association of America has been very active. The website maa.org has details. Philosophy of mathematics that philosophers far removed from mathematics are engaged in is not the same as discussed by mathematicians and philosophers that are at least knee deep in mathematics.

Summing up this entire philosophic exercise, there is nothing like the intellectual life in a US university! It provides an ultimate buffet of stimulating ideas.

April 09, 2005/June, 2014

[PS: Because of some common intersection, a variation of this reflection is also included in Scattered Matherticles: Mathematical Reflections, Volume I (2010)]

COMMENTS

You must be kidding yourself, in case you are of the opinion that I am anywhere near that philosophy. The question is not that Hindu philosophy is divorced from LIFE. The question is why HINDUS WENT IN THAT DIRECTION. Next question is if that direction is causing suffering why are they not changing the direction? Third question is how can we send them in the direction of utility? Or can we send them? My answer is yes. **Subhash Sood, MD**

Very interesting observations and analysis. **Angel Muleshkov, mathematics professor**

A thought provoking reflection. Well written! I enjoyed it. Thanks. **Satish Sharma, Social Work Professor**

Thank you, Satish, for your thoughts. I have some thoughts too about philosophy and the concerns you raise, but we should probably talk them over in person. Thanks.

Steven Rosenbaum (Emeritus Philosophy professor and former dean of the Honors College)

Satish, I appreciate your thoughts and comments on math and philosophy. I'm all for more math requirements in PHI. Most of our majors groan at the thought of taking anything past college algebra (MAT 124). Our chair is quite annoyed with my requiring a prerequisite of a passing grade in Precalculus II (MAT 127) for our PHI 109 (Intro. to Formal Logic). The CSC majors must obtain a "C" in this course in order to take the PHI 421 (Symbolic Logic) requirement. We had a few contretemps about my not wanting PHI majors to take these courses w/o prerequisites. The truth is that w/o a math background they can't do the work. Your other comments are well taken.

Discussions and/or courses about soul, existence of God, etc., are not considered viable topics for philosophy. If it doesn't fit into Hume's two boxes **(Relations of Ideas and Matters of Fact)**, it ain't got meaning!!

Would you ever be interested in giving a paper to our department on these issues? Todd Jones (ext. 54691) is our Colloquium Committee Chair. Thanks again for your thoughts, **Mary Phelps (Emeritus Philosophy professor)**

PERSONAL REMARKS

MATHEMATICS AND HINDU *DHARMA*

[May 20, 2014. The following text is modified from a paper presented at the annual meeting of the Society of Heritage and History of Mathematics held on the campus of the historic Holkar Science College, Indore in December 2005. Some of the ideas expounded in the paper have percolated deep into my thinking. Since then, it was not re-visited until I sudden discovered it in a file. However, it merits its inclusion while keeping the tone of its oral presentation.]

Summary

There is a remarkable similarity between the nature of abstraction in mathematics and lofty flights of imagination in Hindu religion - or Hinduism used synonymously. There are no bars and boundaries in questioning and hypothesizing. By and large, the Hindus love to think 'out of the box'. The paper brings out levels of inquires in both areas. In the context of the conference theme of history and heritage, the paper spells out how the Hindus have made fundamental contributions in mathematics and religious thoughts. This similarity is also reflected in the lexicon of rich mathematical symbols and exotic symbols of Hindu mythology.

Mathematical marvels of the Hindus are independent of the social, political and economic conditions of the nations in which the Hindus live. This is an incredible hallmark of the Hindu religion. Generally, they excel in the areas of 'Pure Mathematics'. It perfectly corresponds with the practice of Hindu belief system that is often so far removed from the nuts and bolts of day to day life.

Background

The genesis of this paper needs a little introduction and as well as few words concerning its background. The theme of this paper has been brewing up in my mind for the last twenty years. Off and on, I briefly shared its core with a few close friends and colleagues. But I lacked conviction to present them publicly. This conference with an emphasis on history and heritage has provided an ideal forum to do just that. I felt it was time to publicly come out of an intellectual cocoon before time runs out on me.

Looking deep into my psyche, I realized that it is connected with a popular political cliché in India of the 1990s, *garva se kaho ham hindu hain* – means, "say it proudly that we are Hindus." I vividly recall the year 1955, at 15 years of age, I was required to fill out an entry on my religion in the college admission form. I wrote with youthful, but misplaced idealism, the word **Humanitarianism** as my religion! I knew in my heart of hearts that I was not a Sikh, not a Muslim, not a Christian. The fact that I still remember this innocuous incident, it goes on to show that a denial of my faith Hinduism clung to the core of my personality.

Yes, it took me fifty years to come out of this shell. With public admission and confession, I feel absolved off a burden that had weighed down on me for so long. In the process, I realized what it meant to be intellectually timid. It is certain that the 21st century will be chronicled as the beginning of the Hindu renaissance.

What is the essence of the nature of mathematics?

It is the most basic question that needs to be addressed. In my opinion as a mathematics educator and student for 50+ years, the **absolute nature of mathematics lies in its abstraction**. Any dichotomy between the so-called reality and abstraction in the world of mathematics and science is very grey. It is like a line in dune sand. In any mathematical system, its axioms, postulates and heuristics are born out of a distillation of the physical phenomena around it. The power of its deductive reasoning leads to results that are called theorems and propositions. They may not be obviously connected with the nuts and bolts of daily living as generally understood to be a common reality.

What is the essence of the nature of Hindu *Dharma*?

The essence of the Hindu ***Dharma*** is its unrestricted and unstructured infinite freedom of thought and ***karma***/action. No Hindu scripture or guru has ever set an upper bound on the flights of mind. Its analogy with a set of real numbers on a line from minus infinity to plus infinity is perfect. There is no starting point associated with any great individual in the hoary past of Hindu *Dharma*. It corresponds to minus infinity! What else could it be? The march of ideas continues with each moment to potentially plus infinity.

Infinitude in Hindu *Dharma*

The number of Hindu scriptures, treatises and commentaries is an index of the relative vastness of Hindu religion. They emanate from freedom to interpret any dogma, doctrine, and personality. The number of Hindu gods and goddesses; *devis* and *devatas* are only representations of ultimate reality. Furthermore, it is reflected in the varieties of Hindu temples, sects, and worships. Consequently, in Hindu religion, any dichotomy between formful god and formless god disappears; and so does it between atheists and theists.

Observation on Hindu Heritage

The names of the ancient *Rishis*, like Vyas, Bhrigu and Vishvamitra, as frequently encountered in the Vedic literature and scriptures are not the names of specific individuals alone. It has been grossly missed in the Hindu ethos. They also stand for the schools of thoughts they captured and cultivated. Great ideas continued to flourish out of them for several centuries. It is no different from modern corporations like, IBM, Ford, GE and Bell. They were founded by individuals, but several inventions are attributed to the corporations for the basic work done by their scientists in employment. With modern mergers and takeovers of companies, the names of the corporations' CEOs continually change that people don't care for them at all.

Some Startling Figures

For a perspective on the role of religion in intellectual development, here is one instance. During the last 50+ years, relative to their population, I have hardly come across Sunni Muslims earning PhDs in the areas of pure mathematics and, even less so, of distinguishing themselves as researchers. The more I became aware of it, the more testing was done on this hypothesis. The Shiite Muslims, Iranians in particular, are exceptions. About eight years ago, I shared this observation with an Iranian colleague who was a bit surprised. However, he added that after the early Arab conquest of Persia – including modern Iran, the Persians did not accept all the tenets of Islam including the Arabic language. This millennium long independence of thought may go some way into explaining the difference.

In 1968, when I went to Indiana University (IU) for an American PhD, it was noted that out of 18 Indian students, fourteen were Hindus, three Christians, one Sikh, and no Muslim! Another interesting fact from my IU years is of women doing PhDs. In 1975, Elizabeth Moore was the first woman in the 150 -year history of IU to finish PhD in mathematics; however, three Indian women had earned their PhDs 3-4 years before she did! Since the 1960s, the number of men and women in India going for PhD's in mathematics are lately seen in the ratio of 20:80. The relative dominance of Hindu women in mathematics is very impressive. There is no better explanation for it except the ultimate freedom of thought in Hinduism.

Still, this data may not be sufficiently large and random enough to establish affinity between Hinduism and mathematics. But it certainly poses an open question, if someone wants to pursue it as a research topic in the fast growing area of ethno-mathematics. However, my data of 45+ years is not a small sample either.

Foundations of Religions

Here, it is not intended to make any studious observations, compare two religions, or pass general judgments. Broadly, Islam is defined by its five pillars: Faith/***Shahada***, Prayer/***Salat***, Fast/***Saum***, Charity/***Zakat*** and Pilgrimage/***Hajj***. Christianity is likewise simple: a Trinitarian Deity consisting of God the Father, God the Son (Jesus Christ) and God the Holy Spirit –within the context of Bible and Church. Judaism and Hinduism are the most ancient religions in the west and east respectively. Any attempt to define Hinduism by **This** or **That** runs into contradictions. In a mathematical jargon, given a belief, there are some Hindus who believe in it and some who do not believe in it. The point is that both groups can coexist peacefully.

When I talk about mathematics, the focus is not on mathematics done for MA/PhD degrees. It is the creation and discovery of new concepts and mathematical modeling. That requires uninterrupted and intense concentration of mind for days and months. Any regimentation of the day obstructs the process of super creativity. For example, if one has to interrupt research for prayers five times a day, the deeper problems of life can never be solved.

Examples of Super stars

S. Ramanujan (1887 - 1920) has left unmatched legacy in Number Theory in the hallowed halls of Cambridge University. Today, Manjul Bhargava (1974 - Present) has illuminated the hallways of Harvard and Princeton. At the age of 28, he was the 2nd youngest tenured full professor of mathematics (Number Theory) in the Ivy League history! In Aug, 2014, he won the Fields Medal, the highest honor in mathematics. In between these two great mathematicians, there have lived hundreds of Hindu mathematicians all over the world who have achieved acclaim for their mathematical creativity.

Does it mean that the Hindus have a fair share of Fields Medals and Nobel Prizes, etc.? The answer is clearly and emphatically, NO. It is also an open research question. However, my explanation is beyond the scope of the article.

It is mainly due to the spirit of free inquiry being their bed rock that Hinduism breathes in them. Hindu scriptures are incredible. The multilayered meanings in the Vedas, metaphorical interpretations in the Upanishads and rarified verses in other scriptures, throughout the millennia, continue to defy any single-track human intellect. They point out to a continuum that corresponds to mathematical works that is enveloping the world in the shape of an inverted pyramid.

For all intents and purposes, Hindu way of life contours an individual to a particular type of thinking. By being fully aware of it, one can optimize one's potential. I don't mean to direct everyone towards mathematics. Nevertheless, every branch of knowledge has its 'abstract' side.

Conclusion

Hindu mind is a contemplative mind; it is a meditative mind. It is relatively not practical. A British officer posted in India wrote in his 1776 travelogue: "The Muslims think and then act; the Sikhs act and then think; the Hindus think and then think." The following **mantra** from the **Ishavasya,** the shortest of all the **Upanishads,** beautifully captures a perfect synthesis of thinking embedded in mathematics and Hindu *Dharma*:

Om, Purnamadah Purnamidam; Purnat Purnamudachyate
Purnasya Purnamadaya; Purnameva Vashishyate
Om, shanti, shanti, shanti

There are many translations and commentaries on this verse, each of which adds a different slant on the vast meaning. In some sense, a complete explanation of the nature of Reality and the entire wisdom of the path of Self-Realization is contained in this short summary. Here is one translation pulled out from internet, as I do now know Sanskrit:

Om, That is infinite, this is infinite; From That infinite this infinite comes.

From That infinite, this infinite removed or added; Infinite remains infinite.

Om, Peace! Peace! Peace!

It all looks, reads, and even sounds like the properties of Transfinite Numbers in mathematics!!

Nov, 2005/Aug, 2014

PS:

This paper was dedicated to the memory of Bhushan L. Wadhwa (1939 - 2004), Mathematics Professor at Cleveland State University, and my friend of 35+ years.

HOLOCAUST AND GODEL'S THEOREM

What a powerful day was it today! It coincided with **Baisakhi,** the beginning of a New Year in India. One half of it was for a lecture scheduled in the afternoon to commemorate the birth centenary of Kurt Gödel (1906-1978), who in 1931 proved one of the most famous theorems in entire mathematics. Ivor Grattan-Guinness, Emeritus Professor of History of Mathematics and Logic (Middlesex University, UK) spoke on, *The Reception of Gödel's Incompeletability Theorems by Logicians and Mathematicians, 1931-1960.*

Guinness detailed how this deep theorem, specifically in the area of mathematical logic, was not given its dues by contemporary logicians and mathematicians till the 1950s. It is characterized as an 'intellectual' holocaust of the Theorem! The western world of mathematics was dominated by the German mathematician, David Hilbert (1862-1943) whose celebrated plan (1920) to formalize mathematics was smashed by Gödel's Theorem. Guinness described Hilbert's influence on mathematics as that of a Field Marshal! Hilbert witnessed the purge of Jewish mathematicians at Gottingen. Gödel's Jewish hereditary is uncertain, but he escaped to the US in 1940 via Russia and Japan.

The other half of the day gave a big surprise. After a couple of hours working on my office PC, I usually step out to straighten my neck, back and hands. Five minute away from my office, UNLV's **Barrick Museum of Natural History** is also a place to check out for new exhibits. Today, on display were 85 black and white photographs. I often observe a collage of paintings or photographs without looking at the names of the artists, titles, or other description.

It did not take me more than 6-7 frames and a few minutes to recognize them as the shots of Nazis concentration camps, though there was not even a single human face in any one of them! Nevertheless, my mind provided the missing images of 11 million people who were systematically exterminated like cattle processed in modern packing plants today.

The pictures, taken from various angles of the camps, are relatively recent and taken by the photo artist, Michael Kenna during 1988-2000. The collection is called, ***Impossible to Forget***. The photo exhibition and an hour long documentary, ***Memory of the Camps*** (Frontline PBS, 1985) are timed with April 1945 liberation of the concentration camps after the Nazi surrender.

The documentary has only April 1945 photos and sound tracks recorded within two weeks of the Allied Forces' takeover of more than 300 concentration camps in Germany and its occupied countries. For reasons unclear, it was after 40 years that this material was turned into a documentary! I thought I had read and seen everything that went inside these camps. But here the scenes are unworldly enough to shake one's faith in humanity.

A human paradox is also presented there. The people living in cities only a couple of miles away from the concentration camps either had no idea what went on inside, or were brainwashed by the Nazi propaganda. Or, they were desensitized by its knowledge that they casually went about their daily business. Human nature could be very revolting!

At the end of the lecture, I remarked. "There is a similarity between what I saw in the museum this afternoon and what I just heard about a systematic neglect of the celebrated Theorem by mathematicians from Germany and other countries." Guinness described in his British 'muffled' accent how the Theorem, published in 1931 in an Austrian journal of repute, was not mentioned in the works of distinguished logicians, Hahn (1933), Quine (1934), and MacLane (1934) - to name a few! Curiously enough, Hahn was Gödel's PhD thesis supervisor in 1929, and had accepted his epochal paper for publication in 1931!

Other famous European mathematicians like Dieudonne, Hardy, and Bell did not include the Theorem in their books and monographs. The US mathematics was not on the international radar in the 1930s! This reminds me of the ***New York Times*** in 1945 not publishing the Holocaust pictures and news reports coming out of the concentration camps! The Tsunami impact of the Theorem was eventually recognized like champagne bubbling out when the cork is removed. Likewise, the Holocaust has transformed the surviving Jews into a nationalistic power in a new state of Israel, born in 1948.

The argument, that in the 1930s only a few mathematicians understood the Theorem, or it was of remote interest, does not hold. The top professionals in the field deliberately just chose to ignore it. The lesser ones only followed suit. The intellectual traditions in the West (different from the Eastern traditions) build upon past references. Few top researchers have the time to understand others' works. Often the inaccuracies perpetuate in textbooks and research papers. At times, research enterprise is no different from the one run by intellectual Mafia.

The Nazi Germany, or for that matter every authoritarian regime, revises aspects of its history. They did not stop at the extermination of the Jews and the dissidents, but also undermined, minimized and distorted their intellectual achievements. There are hundreds of instances. The German Ministry of Education had the fullest control over the intellectuals, professional organizations, conferences, foreign visitors and invitees.

A lesson of history is that once a dictator wins over the intellectuals, then the masses are easily converted, and conversely. There is only a small window of opportunity when individuals can speak up against an 'historic' wrong before it is too late. If that opportunity is missed, then the voices of discord are snuffed and muzzled out ruthlessly. The price of freedom, like in the US today, is to be paid by constant vigilance, willingness to fight, and readiness to go at war.

The denial of recognition to Gödel's Theorem, or the holocaust of six million Jews is the ultimate price a society pays for not participating in its national politics. This is not the first holocaust in history. During the last one hundred years alone, the countries like Russia (under Stalin), China (Mao), Cambodia (Pol Pot) and India (British 1895-1905) have suffered from genocides far worse in magnitudes or brutality. Ninety minutes in the museum and ninety minutes in the lecture opened my eyes again to the dark past, present and future of mankind. **It tells what a single man can do, and what a man can undo!**

April 13, 2006/June, 2014

[PS For its intersection with general reflections, its variation is included in the *Scattered Matherticles*: *Mathematical Reflections*, Volume I (2010).]

COMMENTS

1. I really enjoyed this -- hadn't known about Gödel's being ignored. Thanks. **Tom** Schaffter, Retired Math Logician

2. Dear Professor Bhatnagar, Excellent. This is easily one of the best of your *Reflections* to me as I appreciate each line and each word of it This month's *Notices* of AMS is almost whole of it is devoted to Gödel. The fact remains that the world has yet to grasp Gödel's work fully in its right perspective as you have rightly pointed out.

Personally, the only bit that I have done is his contribution: If Mathematics is consistent without the Axiom of Choice, then it will remain consistent even if the Axiom of Choice is added to it. This gave a complete relief, as we know, to the great headache of mathematicians regarding the consistency of Mathematics itself once for all. I know Grattan-Guinness personally. Hope you have come across the book: 'History of the Mathematical Sciences' which I authored with him. **BS Yadav**

3. Hello Satish, Gödel was not Jewish. His escape from Austria was the result of Hitler's anti-intellectualism, not his anti-Semitism. Gödel's position as a professor (Privat Doszent) was eliminated. As an Austrian, his citizenship (and hence, his German citizenship) meant that he would probably be asked to join the military. Rather than face that, he fled Austria. He seems to have been completely unaware of politics. He seems to have thought of WWII as happening outside of his realm. He waited so long to leave Austria that he couldn't leave by the usual Atlantic route. Hence, he left across Russia, and through Japan, as you noted.

Although I find it surprising that Gödel's theorems took so long for acceptance, I'm not sure I see the connection between the Holocaust and the slow reception. There are numerous biographies of Gödel readily available. Perhaps we could talk about Gödel sometime in more detail. Thank you for coming to the talk. **Ian** (Philosophy Dept)

A CONVERGENCE OF SLAVERY AND MATHEMATICS

"What shall I gain from learning mathematics?" is a perennial question that the students often ask out of curiosity, ignorance, laziness, or confusion. When this question was posed to **Euclid** during his tenure in the famous Greek Academy founded by Pythagoras, he instructed his attending slave, *"Give him a coin if he must profit from what he learns."* Yesterday, after reading this popular historical anecdote in Stillwell's book, *Mathematics and its History*, I stopped to ponder it over.

All great empires are partly built on the sweat, blood and brawn of the slaves. After consolidating the Macedonian empire that Alexander had inherited, he expanded it all the way to the east to the western fringes of India. Depending upon the social conditions, the local population was accordingly subjugated and ruled. Two centuries later, the Romans, in their heydays, continued slavery as the movie *Gladiators* captures it from a sports angle. The Arabs enslaved the Africans and infidels long before the Europeans shipped slaves over Atlantic in the 16th century.

Thomas Jefferson's contributions to the American public life and politics are very well documented. His intellectual prowess was phenomenal. However, he maintained a stable of slaves. A recent movie has brought out his relationship with a black slave girl and fathering kids with her. The historical fact is that slavery is set in the foundations of America. A new empire is built, but it flourishes only, when the society conquers. The dilemma is that the conquest over men is generally a part of the total conquest of ideas in science, art, and literature. The great American ideas and ideals of the 18th and 19th centuries are drawn from the immemorial works of ancient Greeks and Romans!

Well, my second thought from Euclid's quote was that in every age there are individuals who pursue knowledge for the sake of knowledge. The story of Grigory Perelman of Russia, also hitting the internet yesterday, is of a modern Euclid. His declining the Fields Medal and $15,000 associated with it must make Euclid smile in his grave!

No matter how abstract a mathematical, philosophical, or literary work is, some one is directly or indirectly supporting the genius for his daily bread and butter. Some geniuses may not be good at working in teams, or feel comfortable in lime light. However, a stern lesson of history is that in the moments of survival of the society as a whole, the individuals must toe the collective line until the national danger is over.

The controversy of pure and applied knowledge (mathematics) can be traced to the eons of time. Aristotle was highly applied in his work. Early on, he tutored Alexander, and later on, he and his nephew Callisthenes counseled him during his conquests in the Far East. The WWII was first won on the blackboards and in the laboratories of USA!

Aug 24, 2006

GAUSSIAN CHAIR

"Teaching is hindrance in doing research." I heard this quote 30 years ago from a colleague who was then a rising mathematics researcher. Since then, I have tested this hypothesis numerous times, and statistically speaking, it is true in 98 % of the cases modulo what defines a researcher in an institution. This remark is attributed to Carl Friedrich Gauss (1777-1855) acclaimed as the greatest mathematician of the 19th century.

However, last Monday a topical investigation brought another side of Gauss in focus. *"Gottingen in 1846 was not the Mecca for mathematicians as one would have expected it to be with the great Gauss in the chair of mathematics. Professors kept aloof from students and did not encourage original thinking or lecture on current research. Even Gauss himself taught only elementary courses. After a year, Riemann transferred to the University of Berlin, where the atmosphere was more democratic and where Jacobi, Dirichlet, Steiner, and Eisenstein shared their latest ideas."* (Stillwell's *Mathematics and Its History*; p 290) For his varied fundamental researches, Gauss was known as prince amongst mathematicians. Gottingen University, one of the oldest universities in Europe during 1930's of the Third Reich, was completely purged of the Jew faculty - including Albert Einstein!

Of all the variables that go into the making of a good math department, chairmanship or headship (used interchangeably) is the most crucial link. The academic environment of a department is directly proportional to the persona of its chair. The 'Gauss' moment has compelled me to roll back and watch my academic video. Since 1959, as a graduate student and faculty member in India and USA, I have observed and dealt with a total of 11 department chairs. The chairs in the US departments govern with faculty consent; in Asian countries, by and large, it is by the authority from top to bottom. It is a reflection of the political systems, as well as their traditional social structures.

Based on my personal experience, the standout chair is SD Chopra (PhD, Mathematical Seismology, Cambridge, UK). During his headship (1962-78), he built a reputed Math Dept at Kurukshetra University newly started in a new state of Haryana. The state had no intellectual atmosphere or traditions. On the other hand, it has been famous for its martial history in providing the best soldiers to the country! Despite limited state resources and parochial campus, Chopra nurtured the Department in two other specialties far removed from his own. That is one hallmark of an effective chair.

George Springer (1967-70), an analyst, turned Indiana University (IU) Math Dept as Number One in functional analysis (Operator Theory). Besides hiring promising young faculty, he pulled away star operator theorist, PR Halmos (1916-2006) from the University of Hawaii. As a speaker, researcher, teacher, mentor, and author of textbooks, monographs, and articles on math education, Halmos was one of the greatest American mathematicians of his time. However, his teaching style and the way I grew up learning math in India were orthogonal. At IU, there used to be a waiting list for his courses!

Chairmanship and quality research do not go together. It is pertinent to be clear; to be or not to be a chair. Chairmanship has perks - including financial, and opportunities for upward mobility in the administration. But if one does not have interpersonal skills of communication, understanding of personnel and disciplinic differences, then the chair tenure may be rough. Graduate schools train students only for research, provide them some exposure to teaching, but nothing about administration.

Whether in the US school districts, or in the universities, the biggest growth in hi-fi jobs is in the administration. Unless, one has a vision to move beyond the department chairmanship, its chairing may not be a very satisfying experience at the end of the day. A good chair mentors and nurtures a few faculty members for administrative roles while he makes waves for upward mobility.

Oct 28, 2006/June, 2014

COMMENTS

This time I am really disappointed by your observations. For example, how do you compare PR Halmos with, say Irving Kaplansky? Both were contemporary and have died this year. Halmos claims having proved just one great result which concerns the invariant subspaces of shifts of infinite multiplicity, but even that claim is also controversial, see H. Helson's Lectures on Invariant Subspaces where he gives credit to P. D. Lax.

In his biography, he puts himself in the 4th category out of the four categories in which he divides all mathematicians. This too is arbitrary. In fact, the period of his life-time has witnessed tremendous developments in mathematics, and it is difficult to fathom where he stands, but certainly not above the middle. His book on measure theory will be hailed as his greatest achievement. S. D. Chopra was not a mathematician, of course. Whatever traces he left have become invisible and soon be vanished. The Department at KU is horrible these days. **BS Yadav**

I wrote: I love the passionate comments aside from our disagreement on one or many points that I touch upon in one piece. Halmos never ranked himself in the top echelon of mathematicians in functional analysis. It is his sum total of all these that I am evaluating: as a public speaker, researcher, teacher, mentor, and the author of textbooks, research monographs and articles on math education. Let me also add, math expositor in media.

IU did not have a hall large enough to accommodate his public lectures. He was like a Beatle in mathematics! All his books are still in print! Writing a book is not easy, but its selling for 4-5 decades is extraordinary. Publishers loved him; translated his books into many languages! He and his wife donated a staggering amount of 3 million dollars to the MAA five years ago! No mathematician has done it - another mark of his greatness amongst the contemporaries.

Chopra's legacy was carried over by Sarvajit Singh at MDU, Rohtak and Kehar Singh and Ranbir Singh at GND, Amritsar. Eventually, everything vanishes - including men and their so-called great ideas! Keep your ideas bouncing off me. I admire your mathematical vitality at 76. Incidentally, I was a student of both Halmos and Chopra!

PERU (INCA CIVILIZATION): A PERISCOPE

I left Las Vegas for Peru (means the **Land of Gold**), south of Equator, on June 15, 2007, and returned on the 24th. It was my first excursion into the Southern Hemisphere. The journey time of nearly 22 hours was the same for going to Peru or India. Historically, like the US today, India used to be the world destination till the 18th century. Various Muslim nationals like the Arabs, Afghans, Mongols, Mughals and Turks started invading India since the 11th century. European explorers also headed to India - starting from the 17th century when their navigational science and political power developed enough to sail all the way to India. Earlier, when the Europeans ships landed on the Atlantic coasts of America, they thought they had landed on India! That is why all the natives of the Americas continue to be called Indians till today.

First, talking of the naval powers, I often wonder how the Hindus lost it. The Indian states of Bengal, Orissa, Andhra Pradesh, Tamilnad, Kerala, Maharashtra and Gujarat have coastline length comparable to that of the European nations. The Indians ships sailed to Africa, Far East and Middle East through the 4th and 5th centuries, the last golden period of the Hindus in India. Enigmatically, by the 19th century, crossing the ocean meant social ostracization. Gandhi described this predicament when he was to sail for England in 1886. The misplaced emphasis on vegetables deprived the coastal people of their staple protein diet from sea food. Above all, the loss of adventurous spirit for ocean fares made India vulnerable from its porous coastline too.

In Peru, I saw a lot of new land, customs, traditions, and life styles during a week that happened to coincide with annual *Inti Raymi* (means Festival of Sun) of jubilation. I heard different languages; saw colorful pageantry; tasted different foods including guinea pig, alpaca, and llama; enjoyed Peruvian beer, Brahma and Cusquena! It was all due to a National Science Foundation course, *Ancient Inca Mathematics and Culture: Cuzco, Machu Pichhu and Sacred Valley*. Though it was for the college teachers, but a high school teacher and her IT fiancé were also amongst a group of sixteen from ten US states. Seven participants were from mathematics alone.

The decision to leave the laptop was good, as I could unworriedly remain out of the hotel room. However, I took plenty of notes for my ***Reflections***. But it has been more than ten days and I have not cranked out even single one; uncharacteristic of me! In Peru, I was bubbling with ideas. At times, I missed speaking into a tape recorder for its transcription later on. Even notes were time consuming. This ***Reflection*** is a smorgasbord of Peru ***Reflections***; very condensed that a single sentence may be expanded into a nice article. But I don't think I would go back to individual ***Reflections*** as the tides of new ***Reflections*** continue to knock on to my consciousness.

The course was directed by two professional archaeologists with PhDs from University of Texas, known for their seminal archeological work on Mayan Civilization. I also had attended a similar course in 2005, namely, ***Ancient Mayan Mathematics in the Ruins of Quintana Roo, Yucatan Peninsula, Mexico***. The lectures, discussions, and site tours were very stimulating. My thoughts were bouncing all the time from one pole to the other.

Archaeological work is very expensive. It requires millions of dollars in excavation, involvement of various experts for deciphering and dissemination. Any ruins, about 1000 years old, may still look ordinary mounds of earth - often covered with trees and dense foliage. Only good research projects can get research grants and permits from foreign nations. What benefits the universities get in return? Hiram Bingham brought back 70 boxes of material to Yale University after he unearthed the ruins of **Machu Picchu** (means big mountain) in 1911. The national park entrance has his name plate. Also, a tourist train, running between Cuzco and Machu Picchu, is named after him.

Most archaeological research is driven by long and short term goals of applications. The West also values knowledge for the sake of knowledge. It enhances the prestige of the universities to attract bright students from all over the world. Later on, they support their alma maters. Rich universities get richer and more prestigious with new researches and service to the nation through distinguished faculty and administrators.

For example, President Bush tapped Robert Gates, President of Texas A&M University as Secretary of Defense. Henry Kissinger of Harvard has counseled many US presidents. The poor and developing nations are years behind and perhaps will stay back! **Yet, all great civilizations fall and new rise up!**

Another ricocheting thought was that the countries of Asia, Africa, and America were colonized by half a dozen European nations; England, France, Spain, Dutch, Italy, and Portuguese. Germany always aspired to be a supreme power in Europe. During the last 100 years, all the major archaeological projects in the world have been undertaken by only a dozen universities: Oxford, Cambridge and Paris in Europe, Harvard, Columbia, Cornell, Yale, Chicago, Berkeley and Stanford lead in the US. The deep pockets of these universities allow the faculty to re-discover the lost lands and empires that their forefathers had decimated a few centuries ago!

I found this thought really awakening! **Is it now a reverse intellectual colonization**? The well connected young men and women from the poor and developing countries get elite education from great western universities. After completion of studies, they go back to their home countries and occupy positions of power and influence. They implement foreign management and planning models that come out of thinking that had conquered their forefathers!

Let me put the lost civilizations in perspective. For instance, the Inca civilization (1200-1600) comprises a land mass of the size of a typical European country, or a state of India. The study of all the lost empires and civilizations are divided between great universities as their region of influence. Despite collaboration, there is a stamp of being the first. Indus civilization and Harappa Civilization in a region common to present India and Pakistan was accidentally uncovered by a British surveyor while laying out new railroad tracks in the 1860's. Once the British left the Indian subcontinent in 1947, this archaeological work has made little headway. For the Pakistanis, there being no Islamic connection with Harappa and for Indians, so no interest. **Archeological work measures the total development of a nation.**

The Inca civilization was decimated by the Spaniards in a brief period of 36 years (1536-72)! It boggles the mind. Nearly 50% of the populace died of small pox carried by the Spaniards into South America. The native people, having no immunity against the European diseases, died like flies. Though unintended, perhaps it was the first biological warfare waged in South America more than once. The surviving native population was emaciated, ill equipped, unprepared and divided, that it was quickly run over by the Spaniards!

It is troubling! If an individual or nation does not fight to protect its honor, valuables and treasures, then they will be snatched away sooner or later by the barbarians or powerful. It happened to peaceful Tibet in 1950 when the Red Army of China annexed it. Due to divisive caste system and misplaced notion of non-violence, on the top of general weakness, the Hindus lost their will to fight, and hence lost India, their only home land. Either, Incas did not have secret intelligence on Spaniard victory over the Aztecs, 200 miles north, or they just ignored it. The Inca kings did not spend resources on newer weapons and defense. Consequently, they lost all their gold, treasures and lives. **I see these ominous shadows casting spell on the US today that has come under attack from within and without**. The US public, at large, is impervious to these dangers.

Francisco Pizaro conqueror of the Incas was a pig farmer in Spain. His cousin, Herman Cortez who conquered Aztecs 10 years earlier, was perhaps a chicken farmer. I am amazed as similar stories come from India too. Robert Clive, who in 1757 established the English foothold in Bengal, was a clerk who sailed to Canada to seek fortunes. After meeting no success, he was transferred to the East India Company, in Calcutta. Spaniards also remind me of the Turks who used to carve out southern states of India among themselves before invading those regions 4000 miles away. It is disgusting to understand how one nation wins and the other loses everything. History awakens everyone.

There are legends that Spaniards took lot of gold from the Incas. **However, I was curious to know how did the Incas extract the gold in the first place?** How and where they mined the Andes Mountains or panned the rivers, are unclear? Having lived in Nevada and India, I know the engineering skills required in mining gold and silver. The genius of the Incas shines out in the construction of various walls and ramparts with rocks cut and polished weighing from 100 lbs. to 20 tons. No cement or mortar was used to bind the stones. It is a mystery how the Incas built roofs as there are no signs of any woodwork despite thick forests surrounding the ruins.

There is an isomorphic thinking in the Western intellectual tradition since the 18th century. I label it Darwinian, in general. It cuts across all disciplines like, geology, geography, biology, anthropology, psychology, and archaeology. Its underlying assumption is that every aspect of life and living was primitive in eons of time. It is so funny to see the timelines shift by a few hundred to a few millions of years. Also, **the existence, say, of any pre-Inca culture justifies Spanish colonization of the region.** The British disenfranchised Indians by creating myths about Aryan invasion of India, the most preposterous and shallow theory! But intellectual mafia has spread it for two centuries. Its repudiation and refutation are not reported in mainstream!

Finally, I must add some remarks on **Inca mathematics**. The fundamental nature of mathematics is as intuitive as of a dialect in speech, musical notes, or etchings in geoglyphs. As a society organizes into an empire, the role of mathematics gets sophisticated. The Incas having advanced knowledge of mathematics is evidenced by their surviving monuments. Inventory and record keeping are the basic needs of any big organization. Incas developed *Kipus (Khipus)* using decimal system for tracking vast inventories. At the conclusion of the course, I made a *Kipu* of grade distribution of students in a course. While explaining its logic, I deduced that the Incas had a written language too. **It may be lost, but any claim of its non-existence is absurd!**

July 05, 2007/May, 2014

COMMENTS

Bravo for a very informative and sophisticated report. Yes, it's a *Reflection* and a report. It's full of so many interesting observations and facts. Thank you for your profound thoughts and experiences. Fondly, **Dutchie**

Dear Satish: I liked reading your new reflection. I only wish that you should start writing popular articles for magazines. You are good. However, you run away from the hassles of dealing with editors. It is a paradox from a rough and tough person like you. .....

Thanks Satish, very informative. **Gopal**

Hi Satish, Your comments are interesting. The Aztec had a written language. The Incas, even though a very advanced civilization, did not seem to go past a method of keeping records of things in their store houses. It seems written language was only invented 5 times and then the idea spread. Once a Cherokee observed the method of writing from the invading Europeans he quickly devised an alphabet and wrote down the Cherokee history. (I forget the Cherokee's name.) Of course, small pox killed off the Aztecs too or Cortez would have lost his battle as they had him boxed in. It did not help either. They were pressuring the neighboring tribes continually for sacrifice victims, and extorting goods from them. Small pox played a big pole in decimating the American natives living in the Mississippi valley way before they saw a European. It presumably came from natives from Florida that became infected. The purposeful distribution of blankets taken from British small pox victims was an additional factor. Moreover, some USA presidents practiced a policy of genocide quite similar to Hitler's but on a grander scale. (The History of the American People from 1492, Howard Zinn). I can add more details to all of this but it is all very depressing. **Bob Gilbert**

Dear Bhai Shri Satish Ji, Saprem Namaste! Just now I opened my computer & I have gone through your article on INCA CIVILIZATION, but with a cursory reading only, leaving my serious reading of your article REPEATEDLY later in the day. I was wonder-struck, how you could manage to study in a brief period so deeply about all the Peruvian

History, its culture its Heritage, its Archeological ruins & their history which actually gets enshrined in the archeological monuments as was the case in Harappan & Mohan Jodro excavations (which you so rightly compared, totally Hindu Culture, entirely different than Islamic/ Muslim culture of today's Pakistan), the brightness of Peruvians in Mathematics with its wonderful Geometrical usages in constructions of Heavy Stone Monuments & buildings without Mortar or anything, its various Mining & wealth of Gold & Silver & amazing way of how they were recovering these precious metals, their & Spanish Naval powers with comparative studies for each with those of India & Spain in particular and so on.

I was amazed to observe your fantastic Grasping power by just a glance of those Peruvian Sites during your brief visit. Your tremendous knowledge, intelligence & every thing is so far remaining UNTILIZED by our Arya Samaj, which should be used as early as possible for its own benefit, if not otherwise. Arya Samaj should also remember that when the GREAT MAMAMJEE (SWAMI DEEKSHANANDJI SARASWATI) was so intelligent & highly learned Scholar, then why his own nephew is being side-tracked, who requires to be recognized & utilized for the good & benefit of Arya Samaj only.

Possibly, the problem is, your talks or writings are far above or beyond a common person's understanding, they are more Scholastic in nature than ordinary way of putting those things to lesser knowledgeable & of lesser intelligent people. All are NOT as intelligent as you yourself are. I know, for you, it may be very difficult to say in ordinary words, but possibly some of your students at your end, may help you to write in a common & understandable language & manner. Warmest Regards, **Satish Gupta**

Very interesting. I found that trips like these were generally stimulating and often in ways that I had not expected. Looking back, even if a trip was taken for little more than sightseeing or getting away, it usually had the merit of broadening or deepening our interests. As for war and the looting of treasures, I can understand on an economic and cultural level the gains sought by aggressors, but I've always thought it an essentially immoral act in any case. Even so, it seems to me that there is something fundamentally more than it; some kind of collective madness people engage in that even aggressive sports like American football fail to assuage. Ah, the mysteries of the human condition. **Robert W Moore**

Satish, Thanks for the Reflection. I did enjoy looking through the periscope. Sincerely, **Doy Hollman**, Math Prof, Lipscomb University.

A great article!! Thanks. Keep it up. **Vinod Kumar** (Retired Engineer settled in Florida)

[A lesson of history is that if a nation loses its sovereignty to a foreign power, then it loses its heritage, and eventually goes down every thing - including their knowledge of science, mathematics, literature, arts, etc]

WOMEN, MATHEMATICS AND HINDUISM

As I walked into the lecture hall, I was pleasantly surprised to find all-girl audience for my talk on *The Old and New Foundations of Mathematics*, organized by Mathematics Department at the South Campus of Delhi University. It took place two week ago. There were 25 graduate students (postgraduate by Indian nomenclature) doing their master's in mathematics.

In the US, one hardly finds ethnic white/black females majoring in math; forget going for MA/PhD. Most math graduate students in the US universities come from foreign countries – since the 1970s from China, and earlier from India. The data on the Chinese women studying math in China is not known, But, Indian data overwhelmingly correlate math with women. A few years ago, I was astonished to see this Delhi pattern in two other 100-year old Indian institutions; Holkar Science College, Indore and MSU, Vadodara. In both places, at least, 95% of math graduate students were girls!

There is an element of universality about women and mathematics. It is a good research topic for mathematical psychology, and gender and mathematics. My personal data of 10 years of teaching in India, 3 years in Malaysia, and over 30 years in the US, clearly support that the girls outperform boys by at least one letter grade in undergraduate math courses taught during the first two years. In the US, the number of girls suddenly drops in the upper division math courses. Even the top female students in lower division courses do not go for the next math courses. On casual inquiries, their response can be summed up, that **the US girls do not find the pursuit of math gratifying enough**.

In order to engage the Delhi students, I asked them to write in a line or two on their motivation for pursuing master's. The response was uniform: *good at math, math is logical, fun in problem solving*. Only one student was inspired by a teacher. In the context of foundations of mathematics, I observed that every pursuit of an individual is impacted by subtle beliefs in organized religion (modulo a set of measure of zero). Mathematics, being highly organized discipline, is no exception to the influence of organized religions.

The famous theorems in the foundations of mathematics are marked by out-of-the-box thinking. That is where Hinduism reigns supreme. One is free to inquire any abstract thought, challenge any belief, scripture, guru, or practice in Hindu religion (synonymous with Hinduism). Mathematics flourishes in free minds not easily ruffled. Generally, men tend to pull their hair when confronted by theoretical obstacles. Temperamentally, females and mathematics are ideally made for each other.

A purpose of life is all about actualization of one's potential. The odds strongly favor the Hindu mind in the pursuits of mathematics and related disciplines where unrestricted thinking is called for. Let me stress that Hindu mind may be a person of any faith as long as thought domains are not limited.

At the end, I urged Delhi female students to go to USA for PhDs in cutting-edge areas of mathematics, and also learn new ways of free thinking and collaborative researching.

Nov 13, 2007 (India)/July, 2014

COMMENTS

Hi Dr Bhatnagar. Your statement: "My personal data of 10-year teaching in India, 3 years in Malaysia, and over 30 years in the US, clearly support that the girls outperform boys by a letter grade in undergraduate math courses taught during the first two years." is very intriguing.

In that one year that I was teaching Math 100, just before my stroke, I noticed that also; in fact, I've used the same phrase when talking about it. My girl students were always averaging in the B+ to A- range, while my boy students were averaging in the C+ to B- range. I have often wondered about that in the 30 years since then. At the time, the prevailing thought was that boys were 'naturally better' at math than girls, but I didn't really believe that; I thought they would be generally equal, maybe a slight difference in the averages if one particular student was extremely good it could drag his or her gender's average slightly above the other gender's average, but since the classes were around 30 students each and pretty evenly divided as to gender, the difference would be negligible. I was quite surprised at the difference, at how much better the female students were at math than the male students; as you said, one letter grade higher.

I could account for the difference by two theories: (A) that females were 'naturally better' at math than males, or (B) that the stats were erroneous because while the girl population of the class closely mirrored the girl population as a whole, the boy population of the class was skewed because the boys who were good at math would be more likely to skip my class entirely and start out with Calculus. The problem with theory (B) is that maybe the number of boys going directly to Calculus could be so small it wouldn't account for the huge difference.

Another possibility that could affect theory (A) is that my grading technique. I think I had 1/4 of the course grade as homework, 1/4 of it as the final, and 1/2 of it was the small tests I gave at the end of chapter, with the prevision that I would delete the student's worst chapter test (thinking anyone could have a 'bad day' and I didn't want that one day to mess up their grade for the entire course). With the boys, the worst chapter test generally fell at random and wasn't that much off their averages otherwise.

With the girls however, almost all of them were the first chapter test and it was a major disaster - many Ds and Fs. Then they would come by the office wanting to cancel the class. I would go over the test, question by question, trying to get them to give me a logical reason for each step of the question; modeling my words on Socrates' style of making the student answer their own question. At the end of the question, they would always say, "This is so easy, I can't believe that I thought it was hard" (usually those very words!). Then I'd often have them answer the same question using a different order of steps to make them totally sure of the answer. Then the rest of the course, they were involved in each chapter and had high Bs or As on the tests.

After my stroke I was naturally interested in learning how the human brain works. The two sides of it have different functions generally: the right hemisphere being the language controller, and the left being more spatially oriented. The higher mathematics seems to be less about 'numbers on the page' and more about 'pictures in your mind' (delta-epsilon convergence, etc.). Most brain researchers seem to think that male brains are strongly

dominated by one side or the other, while female brains are much more equal from one side to the other and both work in a more harmonious relationship. That could account for a deeper understanding of mathematical concepts by female students in general than by male students.

I don't really see too much of a connection between theology and math, although I personally use logic to get a handle on documents: their dating, how relevant that is to other documents, et cetera. **Robert**

MATHEMATICS, SCIENCE AND RELIGIONS

Lately, I have been getting wonderstruck by the impact of deeper religious beliefs on the mathematical achievements of particular individuals and societies at large. Often one probes into a mathematician's mind, as to how he/she came up with a breakthrough idea. However, it has not been collectively investigated, say, for any religious group. This area of ethno-mathematics lies in the domain common to meta mathematics and neurology of genius. French mathematician, J. Hadamard (1865-1963) did pioneering work (1954) in this direction.

This socio-psychological problem is an outcome of my mathematical experiences over the last 45+ years. It clicked during the academic year, 2006-07, while teaching graduate courses on history of mathematics and survey of mathematical problems. It is clear that the correlation coefficient of mathematics with major religions of the world is not constant. It varies with each organized religion and its historical chronology. My students, also fascinated with this question, researched a few solid facts and figures. This **Reflection** briefly touches the subject, but it does point out at various directions for further investigation.

Starting with Christianity, its major reformation was the emergence of Protestantism in the 16th century. It subsequently played a key role in the development of science and mathematics. It was a quantum leap in intellectual freedom. Briefly touching the religious depth of a few prominent mathematicians of the beginning of the modern era, Pascal (1623-1662) was converted into Jansenists, a fast growing sect within Catholic Church. His Provincial Letters, promoting Jansenists idea are considered classic in French literature. Newton's (1643-1727) interest in alchemy and Christian philosophy during his later years was driven by his religious beliefs. Euler (1707-1783) was the son of a Protestant minister and studied theology too.

Giving an American touch to the subject, Harvard University, started in 1636, as a religious college, has its present Divinity School, which is as famous as its Law and Medical Schools. The University of California is named after Bishop Berkeley (1685-1753). The US has numerous colleges where science and mathematics are studied along with Christian theology. Mathematics and religion nurture each other in a subtle manner.

In India, the advent of Buddhism and Jainism over 2500 years ago are major milestones in the reformation cycles of Hinduism (used synonymously with **Sanatan Dharma**/Hindu Religion). The resultant emphasis on logic and experimentation generated new sciences and mathematics. It dawned a golden period (400-700 AD) of unprecedented prosperity in every walk of life. As a matter of fact, India became a destination of fortunes seekers, scholars and adventurers. It was like the USA has been a demographic magnet of the world for the last 50 years.

However, the great Hindu mathematicians - like Aryabhatta (476-550), Bhaskara I (600-680 AD) and Brahmagupta (598-665) worshipped Brahama, Shiva, Vishnu, or Ganesh - the major deities in Hindu pantheon. The legendary Ramanujan (1887-1920) worshipped Namagiri, a family goddess. He credited his incredible theorems to her. He often said, "An equation for me has no meaning, unless it represents a thought of God." The seemingly unworldliness and complexities in Hindu religion and its mythology often run parallel to the rarified heights of abstraction and exotic symbols in mathematics.

Nevertheless, it is simplistic to think that religion alone determines the development of mathematics. Stable political systems and general prosperity of a state play significant roles in the growth of science and mathematics. In the class, we examined these aspects under 'the necessary and sufficient conditions' for the flowering of mathematics at every level.

Two startling data emerged, when all the winners of Nobel Prizes, Field Medals, Abel and Nevanlinna Prizes in sciences and mathematics were examined. Let me add that the religious affiliation was derived from the names only. Thus, there is a room for error, the quest is of its statistical significance. The number of Muslim winners was only two, Hindus five. However, the list of the Jewish winners ran into two out of nearly four pages!

It is to be noted that amongst the Muslims a vast majority of scientists and mathematicians are Shiias, an offshoot of Islam emerging during its infancy. However, it does not have any semblance with reformation in Islam. The branching of the Shiias came off due to conflicts in the hereditary rights of the followers of Prophet Muhammad.

In the contemporary Muslim world, the Pakistani Imams (heads) of nearly 100,000 mosques, running *madrassas* (Islamic religion schools), overwhelmingly rejected the introduction of science and mathematics in their curricula. It happened 3-4 years ago, when the Bush administration offered nearly 50 million$$ for the curricular changes. The US administration thought that the study of science and mathematics along with Quoranic instructions may take the Muslim youths away from the path of terrorism. As it is, the *madrassas* are the Harvards for turning out human fighting machines.

After flourishing for six centuries in India, Buddhism, with royal patronage, spread in Afghanistan in the west, Sri Lanka in the south, China, Japan and other countries in north and Far East of present India. The 'reformation' and the spread of Mahayana Buddhism (Hinayana; later on, Theravada) catalyzed eventual development of science and mathematics during the Han Dynasty (3rd BC to 3rd AD) of China.

Interestingly enough, there was a transfer of mathematical applications in astrology from India to China. The knowledge of geometry, in the *yantra* in divination, was transformed into Chinese magical squares (cousins of Latin Squares). Similar influence on the Far East countries is an open question. The whole subject is very fascinating!

Nov 18, 2007 (India)/July, 2014

STIMULUS OF THE FIRST LOVE

"...I am at last getting some time to learn some mathematical physics that I didn't get a chance to learn at the right age!! Better late than never. Of course, my particular interests come with a price, viz., not many people who are interested in it, so that the journey is somewhat alone...."

Today, these lines hit me like running into the first love after 50 years! It transported me to my college days in India of the 1950s. On comparing with the US curricula, my bachelor's and master's were essentially in mathematical physics. Of course, it was all theoretical – no labs. Most of my high moments in 'mathematics' are associated with statics, dynamics, and hydromechanics. Nonetheless, electricity, magnetism and vectorial mechanics gave me some unforgettable low moments.

Raja and I, educated in similar educational systems in India, earned our math PhDs in the US. He is now 75, and retired. We first met 23 years ago while attending a Chautauqua course in *Elementary Particle Physics: Current Perspective,* organized on the campus of famous Argonne National Laboratories, Chicago. However, my interest in mathematical physics did not go beyond popular science books, magazine articles, and occasional colloquium lectures in Physics Department.

In the US universities like UNLV, mathematical physics has lost its appeal. By and large, science has gone experimental, and thus spectacular for attracting college students. The **Large Hadron Collider** is a super technological lab ($10b) ever put together! It is baffling that a 17-mile ring accelerator is needed to understand a tiny photon, its collision with another photon, and more!

The interpretations of experiments are based on measurements and mathematical modeling. Both are prone to errors and subjectivity, and hence new tools of research. On looking at the history of sub-atomic particles over the last 50 years, one characteristic that stands out is that the properties of subatomic particles change as their size becomes smaller. Below a certain threshold, properties both of a particle and of a wave are displayed! At nano scale, 10^{-6} in measurements, physics, chemistry and biology all merge together. However, mathematics remains a tool and language.

Here is my dilemma. In contrast, properties of real numbers do not change with their micro magnitudes. The same real numbers continue to describe subatomic particles! In the derivations of many elementary principles of light, heat and mechanics, I recall using the arguments like 'the product of two or more small quantities is set equal to zero'. This scenario has stood out in my mind. I used to get gutsy feelings of deriving my own principle one day! Well, this feeling is universal and it prevails after having grasped the fundamentals of any subject.

Do we really need algebra of small numbers to understand small particles? One may raise a similar question about large numbers used in astrophysics. Apart from real numbers, science is explained by complex numbers, various hyper numbers, quaternions, tensors and other 'exotic' mathematical objects. A few years ago, my proposal to develop a major in mathematical physics did not take off, as there is no second person in mathematics and physics interested in it! However, going back to the first love has its own rewards. **It is a Viagra of life at 70 plus!**

Sep 07, 2008/July, 2014

COMMENTS

Satish, I enjoy thinking about the mathematics that underlie physics but think of myself more as a dilettante than someone versed in mathematical physics. Although I enjoy "improving" my skills and trying to impart some of the skills and enthusiasm to students, I don't think UNLV has a pool of students who would select mathematical physics as a major. And if there were such a pool, I don't think my background is deep or broad enough to bring them up to speed to leave UNLV as well prepared as I would like.

Meanwhile, I enjoy making little discoveries and trying to improve my overall comprehension of the mathematics underlying much of the physics we do over here. For example, I just learned that you can think of the differential "volume" as a tensor that takes little vectors and turns them into a number -- the differential "volume" of an element. I found this cool because it gave me that best explanation of how the Jacobean arises in connecting the elements in different coordinate systems. Weird about the email. Take care. Ciao, **Len**

BRUSHING OVER THEOLOGY AND MATHEMATICS

"In 1669, Isaac Barrow, the first occupier of the Lucasian Professor of Mathematics at Trinity College, Cambridge, resigned to devote himself to theology, and Isaac Newton was appointed to the chair on Barrow's recommendation." Yesterday, as I read this sentence in Stillwell's book, **Mathematics and its History** (2002), it unleashed a stream of dormant thoughts.

My study and admiration for Newton's life and work has been ongoing. Partly, it is due to the colonial connection between England and India. For that very reason, there was little exposure to other European minds. Before independence, the indigenous sprouts of modern science and mathematics were rare. Despite having read numerous accounts of Newton, there is a time and space when the mind is more sensitive to the depth of a certain idea connected with that era.

A gamut of thoughts that ran through my mind is: Did Barrow find mathematics intellectually less challenging than theology? Did he go for more money, power or/and prestige than the new Lucasian Chair provided him? Certainly, Barrow's study of theology and mathematics must have been close that he could easily jump from one area to the other. Like this closeness between Christian theology and mathematics, is there similar closeness between Islamic theology and mathematics, Hindu theology and mathematics, Buddhist, and so on?

As my mind was still engaged with these questions, a colleague ran into me. When I shared this line with him, he remarked, "Fundamental research in mathematics being more difficult past the youthful years, often mathematicians turn towards soft intellectual pursuits, in later years." Yes, that is a well established since Cambridge mathematician, G. H. Hardy laid it out, in his classic book, *A Mathematician's Apology* (1940).

In a general vein, during 2008-Beijing Olympics, it was revealed that female gymnasts prime up at age 12, even before teen years. If an athlete runs a mile under 4 minutes at age 20, that does not mean it is the end of life if he can not do it at 30. **In academia, there is a misconception or misplaced expectation that one is supposed to stay 'hot' in research all the time.** In professional life, my latest role model is John Madden, who was a reasonably good football player in school, college and NFL, a far better coach in college and NFL, and perhaps one of the best color commentators of football today. He transitioned from one phase of football life to the other while setting new benchmarks.

However, research on Barrow turned out fascinating. He was only 36 when he resigned from the Lucasian Chair! During his Lucasian tenure, he published two mathematical works of great learning and elegance, the first on Geometry and the second on Optics. About this time, he also composed his *Expositions of the Creed, The Lord's Prayer, Decalogue, and Sacraments*. Some of his later works in theology, particularly on the power of papacy, are immortal.

Whereas Newton's statues and portraits are displayed far beyond the corridors of Trinity College, Barrow's statue sits right inside the Trinity Chapel. Trinity became a leading confluence of theology, mathematics and sciences. Incidentally, Astro-mathematical physicist, Stephen Hawking (since 1980) is the 16th holder of the Lucasian Chair. History of Lucasian Chair rightly reflects on the history of British Civilization!

Incidentally, before Newton became the Master of Mint in 1696, his interest in science and mathematics were weaned away by theology and esoteric pursuit of alchemy, the 'art or science' of converting base metals into gold. Newton also wrote several scholarly tracts in theology during 1690's. Alchemy is still 'practiced' in India to dupe gullible people out of their gold. Since the atomic structure of every element is now fully known, one may theoretically pose the conversion question between any two elements. However, the process may be very complex in setting-up and time. The money involved may be phenomenal, relative to the return. For years, I had saved a copy of Newton's hand written paper on alchemy. A student had zeroxed it from the original preserved in the US Library of Congress.

In the backdrop Newtonian era, present off-and-on controversies between creationists and evolutionists are intellectually puerile. As compared with 17th century Europeans, the intellectual making of a person today is lopsided. Generally, the study of science excludes religion, and conversely. In the US state universities, theology is out of the curriculum. To me, it is as ridiculous, as during my student days in India, when one was not allowed to study science and mathematics with philosophy and humanities.

Years ago, psychology dept chair and I were neighbors. One day, I asked him if there was a psychological explanation of human soul. Promptly, he replied, "Soul is not in the domain of study in western psychology!" How can one simply cut out an open inquiry on intellectual grounds? Harvard University started as Harvard College in 1636, primarily focused on theology. Though its mission has broadened since then, but its School of Divinity remains the most famous. The great ideas in science, math and theology have been fermenting in the crucibles of Harvard for the last 100 years with their worldwide impact.

An offshoot of theology is the role Jesuits friars played in modernizing education in European colonies during the 19th century. Some of these celibates taught in several Catholic colleges like St. Stephen and St. Xavier in India. Their high academic standards, personal and professional devotion with deep Christian values influenced generations of Indians. In Europe, from the 16th to 19th century, many discoveries in science and mathematics came out of men steeped in Christian theology.

I have been curious about Islamic theology ever amalgamating with science and mathematics. Five years ago, the Islamic clergy in Pakistan rejected Bush Administration's offer of nearly 500 million dollars for introducing science and mathematics along with Quoranic instruction in over 50,000 madrassas, Islamic religious schools. Without a good nursery, there is no plantation; without a lab there is no science.

The dominance of Jewish scholars in science and mathematics strongly suggest a correlation with Judaic theology. The Nazi Holocaust has pushed the Jewish theology in every aspect of Jewish life. The Jewish Nobel Laureates almost outnumber the rest of the religions combined! In academia, their presence in mathematics is predominant. These achievements are simply amazing for people with worldwide population of only 15 millions, spread in 50 countries!

The question of Hindu theology and its interaction with science and mathematics is uniquely different. The religions of the west stress upon the organization over an individual. Hindu religion as practiced since the 10th century is individual oriented, and it grants the individual absolute freedom of thought and inquiry. The historic Christian crusades against unbelievers, and Islamic Jihads and Fatwas against the infidels, have no counter parts in Hinduism.

Conceptually, the abstract mathematical inquiries and Hindu theological thoughts are the closest. Great Hindu mathematicians of the yore, or the recent ones, like Ramanujan, attribute and dedicate their mathematical discoveries to their personal/family gods or goddesses. Ramanujan's famous line is: "**An equation for me has no meaning, unless it represents a thought of God**." However, the modern history of the Hindus in science and mathematics is relatively too short to make a strong connection with Hindu theology. In India, the Hindus have yet to start self-governance, though freed from the British rule in 1947.

Buddhism, started as a 6th century BC reform movement of Hinduism, eventually crystallized its theology on a far greater analytical plane. It puts equal emphasis on organization and individual. As its ethos spread over China and Japan between 3rd BC and 3rd AD, it transformed the intellectual culture of the new lands while enriching the native traditions. During Han Dynasty, for example, science and mathematics showed first great signs of development.

In the study of mathematics or theology, one easily accepts the premises in one over the other. Without any common platform for the resolution of apparent differences, in a US culture, one out rightly rejects the other. This semester, while teaching a course, *Survey of Mathematical Problems I* (MAT 711), we critically examined the earliest foundations of Euclidean Geometry in 23 definitions, 5 postulates and 5 common notions in Book I of Euclid's *Elements*. Volumes have been written on them. The current foundations of mathematics are no less questionable, and they will continue to be in flux. Yet, the applications of mathematics have been producing marvelous results.

On the other hand, the lives of the like of Gandhi and Mother Teresa are grounded in theology; here, one in Hindu theology and the other in Christian. They have equally impacted mankind. Such men and women live in every age and culture. It is appropriate to ponder over what the greatest mathematical-physicist of the 20th century, Albert Einstein, said about Gandhi, on his assassination: "**Generations to come will scarcely believe that such a one, as this ever in flesh and blood, walked upon this earth**." Einstein combined science with humanism.

In conclusion, theology, science, mathematics, and all other intellectual ideas emanate from a human mind. After ricocheting and traversing, they bounce back onto the human mind. It is wonderful scenario of unity in diversity, and diversity in oneness! The ultimate purpose of theology or of mathematics is to enhance our understanding of the universe, solve human problems, and leave the world a better place for the next generations. Any differences are superficial, indeed. Nevertheless, mathematics remains a 'sport' of the young, and theology of the grown-ups. The twains meet at a Golden Plane!

Oct 18, 2008/July, 2014

COMMENTS

1. Dr. Bhatnagar. One cannot encapsulate reality into mathematics anymore than the reflection of the moon in water encapsulates the moon. Theology is a pointing to the moon while mathematics is a pointing to the reflection. Theology and mathematics harmonize when the winds of the mind are stilled, for it is then that the reflection is clearest and undistorted. Theology has been handed down to us by those who have seen the moon, and not that which is reflected. For those who have handed down theology to us, the ultimate purpose of theology is not to enhance our understanding of the universe, solve human problems, or leave the world a better place for the next generation. The ultimate purpose of theology can only be found in the moon, and not its watery reflection. All of existence is but one thought in the mind of God, non-existence another. These are but two in the infinite Mind. Good luck with your thesis! **Kevin**

    I wrote: I am glad it clicked on you. My thesis is the ultimate unity of all thoughts! I know you are on a path of self-discovery.

2. Satish, that was one of the best of your always interesting essays. A lot of ideas for further exploration. This is not an area I give much thought to because I have little respect for religion and its impact on the world, but that is a different issue. Take care. Ciao, **Len**

3. Hello Satish; Interesting questions. Some stray comments. I feel at the heart is the quest for "The Truth". Theology and Mathematics (along with Physics, Philosophy, Cognitive Science) claim to study the ultimate truth. Though in each discipline it is the theoretical, as opposed to the applied, aspect which chases the Truth.

    ....But what bothers me is some other question. What exactly is Islamic Math? Math based on Quran? Or Math done by somebody who is a Muslim? What is calculus? Christian because its origins from Newton and Leibnitz? Creationism is Christian as its source is their Holy Book. It can be considered Islamic for the same reason. Then what is Evolution? Christian, because Darwin was a Christian? Hope to see you in December. **Pritpal**

4. Dear Satish: Excellent article! This has been an area of my interest. My internet article CHRISTIANITY AND TECHNOLOGICAL ADVANCE (www.icr.org/article 374) is related to this. It touches on Islamic theology and science showing why the growth in science was stunted because of certain limitations in its theology. Mathematics, since it explores the organization of quantitative facts of the universe and also the logical derivation of conclusions, makes use of the theological world view. I have on my files much more detailed information on the theology-science connection including Hindu theology. Perhaps I can share it another time……..

    I will be going to India in November and spending time with a top math professor from India (Brahmin background) who, while in San Diego, shared with me about his deep commitment to Christian theology. Please keep in touch. **TV**

5. A beautiful piece. I have been frustrated with intellectuals who have amputated lines of thought as either inferior or not worth pursuing based on dogmas. I would encourage you to pursue this theme in at least couple of more articles as this is your forte. Also, I would be interested to hear your analysis on "Vedic Math" - specifically how could one culture come up with something so fascinating and then totally lose it. **Harpreet Singh**

6. Hi Satish; It very thought provoking. You have the talent to do so. My engineering back ground is not well equipped to do so. Regards. **Matt**

7. Hi Satish, Mathematics are based on "known" facts whereas theology is based on "known" faiths. As people mature, they learn that the facts in fact are the faiths. Hence this migration from mathematicians to theology is understandable. Though when viewed superficially it appears paradoxical. After all, both are engaged in solving the mysteries of the same universe. **Gopal**

8. I always believed that God is a mathematician of a very high order. He used some very advanced mathematics in constructing this universe. So, Mathematics and theology can go together. St. Augustine says that the world was created in six days because 6 is a perfect number. He also said numbers are the link between humans and God. Shakespeare said in one of his plays there

is divinity in odd numbers. Was he perhaps picking up on the Trinity and the mystic number 3?

I do not know if you have read the book of Revelation from the Bible. It is full of numbers and of number mystics. For example, Revelation Ch. 1: 20 says, "Here is the secret meaning of the seven stars which you saw in my right hand and of the 7 lamps of gold: the seven stars are the angels of the seven churches and the seven lamps in the seven churches..and the famous passage from Revelation, "...anyone who has intelligence may work out the number of the Beast. The number represents a man's name and the numerical value of its letters is six hundred and sixty six"..Innumerable computations of the second coming of Jesus have been carried out... the idea that the end of the world can be computed is quite well known...Anyway... an interesting area for research.. Cheers **Abraham**

9. Hi Satish, Thanks for sending me your Mathematical Reflections. Keep it up. **Jim**, Chair, Department of Mathematics, IU.

HISTORY, MATHEMATICS & HERITAGE

Remark: This article is a modification of the invited talk presented at the national conference of the **Society for History of Mathematics**, held during Dec 19-21, 2008, in Imphal, the capital city of Manipur, an eastern state of India.

Introduction

The title, *History, Mathematics and Heritage* forms a golden braid; or speaking mathematically, a golden triangle, but not a Bermuda Triangle. Mathematics is deliberately sandwiched between History and Heritage. The theme of this year's conference, **History of Mathematics**: *Its Role in Science and Society*, threw a challenge that most presenters seem to have ignored.

History, in contrast with mathematics, is often appreciated in midlife. The makers of history are seldom its regular writers and readers. History, in all its facets, measures a nation's total development. Mathematics is generally studied independent of politics, humanities, and literature. History of mathematics is a patch on a larger history quilt of a nation or society.

This conference is taking place under very abnormal circumstances in India. A couple of times while in the US, I inquired whether it was going to be cancelled. India looks very different from 20,000 KM. A very small number of Indian participants outside Manipur, and I, the sole invitee from abroad, indicate political undercurrents. Therefore, at the outset, I must congratulate several teams of local organizers as well as the national committee of the Society for their faith in the mission and courage to hold it in Imphal. Already, many fond memories have been formed. This venue has stirred a few latent thoughts. In March, 1987, while visiting Jaisalmer (Rajasthan), I was hardly 20 KM from the western-most border common with Pakistan. Today, Imphal may be equally the same distance from the eastern-most border from Myanmar (Burma).

Mathematics and History

History of mathematics is only a small streak of history, in general. One must remain fully awake to it irrespective of one's pursuit in life. Unfortunately, sense of historicity and historical perspectives have never been seeded in Indian schools and colleges, particularly for those studying science and engineering. Therefore, it is incumbent to have its quick overview in the present context.

Following are a few observations on the nature of mathematics in the context of its history:

(i) Mathematics, though man-made, is very unique. It is free of contradictions, but has seemingly contradictory characteristics.

(ii) Mathematics is a favorite discipline of the poor, but it only flourishes in affluent societies, when it comes to fundamental work of the caliber of Abel Prizes or Fields Medals. In the US, I am often asked, "How come you studied math?" "Science being expensive, I had no access to it in my hometown, Bathinda, during the 1950s. Math even costs less than humanities," has been my quick response.

(iii) Yes, mathematics is done with paper and pencil, but recent breakthroughs are coming with computing power.

(iv) Mathematics may appear to be a passion of lonely persons, but bigger national problems are tackled by teams of mathematicians. The present US mathematics epitomizes collaboration and interdisciplinary approaches. In Indian context, Ramanujan (1887-1920), at Cambridge University (UK), provides a perfect example of a lonely mathematician and team player lead by G. H. Hardy. Littlewood joined them later.

Mathematics and Hinduism

However, we have to ask again and again: **Is there really a history of Indian Mathematics?** Nay, first comes the question of history of Hindu Mathematics. I may have touched an intellectual nerve here. However, it is done with intellectual conviction of a 'neurosurgeon'. While mathematics is beyond religion and national boundaries, but mathematicians, being human, are greatly influenced by them!

Let me expand this thought a little bit. In a jargon of mathematical writing, I statistically claim that **Hindu mind is closest to mathematics**. Because, Hinduism gives an individual the ultimate freedom to hypothesize, question, challenge any precepts, scriptures, and teachers. While using the term Hinduism, I am not nit-picking with terminologies like *Sanatan Dharam*, *Hindu Dharma* etc. The Muslims have known for centuries, as to who the Hindus are; Christians know who the Hindus are, and so on.

It was the result of such open-minded questioning of Euclid's Fifth Postulate for centuries that led to the creation of non-Euclidean and other geometries. In this process, geometry changed in complexion; from visual, it became analytic, abstract, and axiomatic. It is due to open investigation while staying consistent and not letting contradictions come into play.

Hinduism has no place for any kind of Christian crusades, Muslim *jihads* and *fatwas*, and Sikh *hukamnamas* against disbelievers and non-conformists. No judgment is intended here. However, I do pose an open question for the historians of mathematics: What are the necessary and sufficient conditions for the flowering of mathematics; be that mental, political, economic, or social?

My thesis raises significant research questions in a growing area of ethno-mathematics. If Hindu mind is so good at math, then how come they are not proportionately represented in the top echelon of world mathematicians? At the same time, the number of Hindu students doing masters and PhDs in India and abroad is far greater than their demographic proportions. During the last four years, I have checked this observation during my visits to Delhi University (South Campus), Holkar Science College, Indore, and MSU, Baroda - all more than 100 years old.

The essence of mathematics lies in its deductive nature, or in a mathematical terminology, in its sequential connectedness. That is where the organizing nature of mathematics comes out. If mathematics organizes minds, and Hindus are good at mathematics, then Hindus must be fully organized, or organizable. **Isn't that a paradox from recent history of the Hindus?** This statement is worthy of research in mathematical psychology and history of mathematics.

Medieval Hindu History

Going back to the question of history of the Hindus, before the 11th century, the present usage of the terms, Indians and Hindus were synonymous. But the landscape has been changing rapidly since then. Today, Indians include approximately 15% Muslims and 3% Christians. Buddhism, Jainism and Sikhism are offshoots of Hinduism, or the reformist movements in Hinduism. That points out to the dynamic nature of Hinduism.

It may be added that there are Arab histories giving Islamic perspectives on every facet of life including science and mathematics. Since the colonial dominance of ¾ the world by European nations during 17th to 20th century, all histories are written from the stand point of Christian superiority and vintage points.

Simple questions like - who cares about history, who writes history, which makes history, who reads history, and who researches into history, are never to be taken lightly. They influence generations to come. The Hindus have to adopt and **adapt** to western historical outlooks in life, as they have done to modern science and technology.

The Role Intellectuals in National Crises

It is time to put Mumbai Attack (11/26/2008) in perspective, as it is one of the 17 Islamic terrorist attacks in India during the last 5 years alone. It must ring in Mohammed Gori's (of Arabia) 17 attacks on India in the 12th century! However, I take you on a historical roller coaster. When the Roman barbarians had conquered Greece and ransacked the famous Pythagorean Academy, Archimedes, the great mathematician and scientist, was absorbed in his researches. He was making some observations of the Sun when a Roman soldier, walking by him, blocked the sun. Archimedes yelled at the soldier to move away. He was quickly slain by that soldier who had felt insulted by an 'unknown' old man. As a matter of fact, it was the sunset on great Greek Civilization. The sun rose for the barbarian Romans in 3rd century BC, and it lasted till 2nd AD.

Alexander, the Great, had heard of vast Hindu kingdoms, and understood their divisiveness too. He was keen on meeting the great preceptor of the vanquished Pouroosh (In Greek, Porus). Alexander was amazed that the great counselor of the King lived on the outskirts of the town, disconnected from the political realities that had fallen upon his kingdom. Since then, the western frontier of India has remained porous and vulnerable to hundreds of foreign invaders, marauders, and adventurers from Central Asia and Middle East, who have changed the cultural face of India beyond recognition.

My thesis is, that if Indian intellectuals, including mathematicians, do not stand up to present political realities, they may meet the same fate that the Kashmiri Pandits have met several times during the last 400 years. It also happened to the great centers of Hindu learning, Prayag, Ujjain and Varanasi after the Muslims took over these regions in the 12th century. Some scholars escaped to Deep South and some to the Far East.

Research Questions on History of Mathematics

No Hindu nomenclature has numericalized names, like Gupta Kings I, II, III, or mathematicians Bhaskar I and II, Aryabhatt I, II. This is one of the British intellectual conspiracies of miniaturing and distorting facts. It reminds me of the last Hindu Emperor of Delhi, Samrat Hem Chandra Vikramaditya merely referred to as Hemu in history textbooks. My limited research in this part of history of mathematics indicates that Bhaskar, Aryabhatt, and Bhrigu are the names of great schools founded by them; like that of Pythagorean in Greece in ancient times.

It is an exciting research area to establish a connection between the 15th century mathematicians like Madhav and three centers in north. Mathematical activities nearly extinguished in India by the end of 12th century. Whereas, some records of Madhav and Nilkanth survived, but to the best of my knowledge, old Eastern schools of mathematics, in safe havens like Imphal, are yet to be discovered.

For instance, King Dharam Pal (of Pal Dynasty), during his rule in the 1150s, had invited several Hindu scholars from Kannoj, Mithila and other places to his kingdom. It turned Guwahati into a great center of astronomy by the 15th century. Later on, the city came under attacks from an Afghan ruler of Bengal. They destroyed the famous Kamkhaya Temple of Guwahati, as another Afghan ruler Balban had earlier obliterated the famous Mahakal Temple of Ujjain. Such investigations require teams of mathematicians, historians, archaeologists and anthropologists. However, this happens in times of peace and prosperity of a society.

Examples of Mathematicians in Action

Mind, nurtured in freedom, has defining characteristics. During 1960's, the US mathematicians protested against the US government's role in Vietnam. For instance, well-known Indian mathematician, S. G. Ghureye, Indiana University Math Dept Chairman (1962-68), left USA, as a protester in 1968, and settled in Canada. In the 1980s, the US mathematicians protested against the USSR for not letting emigration of the Jews. Incidentally, they have not done it against the US counter attacks on Afghanistan and Iraq. They understand the political realities.

Recommendations

The academic future belongs to interdisciplinary expertise. Therefore, it is time to integrate history of mathematics with economics, political science, literature and sociology, when examining it. Joint faculty appointments foster cross fertilization of ideas and methodologies. The mathematical culture in most US math departments is intolerant towards mathematics education and history of mathematics.

In some US universities, mathematics education has found niches in colleges of education, where history of mathematics courses are offered to prospective school and community college teachers. UNLV Math Dept offers two courses. India has an advantage that in upcoming universities, new faculty lines and curricular experiments can be undertaken.

Climax

Finally, the paper was concluded with a *Sutra* (pillars) on the making of a society based on three strong principles. They are *Shaastra* (Heritage), *Shastra* (Weaponry; the courage to wield power of pen or sword for right convictions), and *Shatru* (Enemy; within and without. Nations fall from within first). This *Sutra* really puts everything in present context and perspective.

Dec 26, 2008 (India)/July, 2014

PERSONAL REMARKS

ON AN HISTORIC IF 'N BUT

A certain question-cum-insight bolts out when the mind and the locale are synchronized. It is also called the right chemistry or perfect alignment of the planets in a constellation. This seems to have happened today, when I asked Ahmed Yagi, **"Do you think that the world's greatest library, in Alexandria, would have been spared from total destruction in 641 AD, if Prophet Mohammed were alive?"** He died in 632 AD, only nine years earlier.

Ahmed, a native of Sudan, microbiology PhD from Helsinki (Finland), taught in Saudi Arabia before moving to Oman two years ago. He is also an Assistant Dean at the University of Nizwa (UN). As I gave him a copy of my *Reflective* article, he shared a horrific story of his library visit in a Saudi Arabian university. During conversation, one thing leads to the other, and I said, "The students, graduating from the UN, after 5-6 years, may not have even single textbook in their homes." The instructors issue the textbooks at the beginning of a semester and collect them back at the end. The textbooks are not sold on or off campus. It does indicate of the low esteem in which the books are held here. Also, apart from 3-4 tiny stationery shops, I have not encountered even single bookstore, or a public library in Nizwa, the third largest and historic city of Oman.

A context to the above piece of conversation is that a couple of hours prior to our meeting, I was preparing for my *Number Theory* lecture. In David Burton's textbook, a passage read, *"For nearly a thousand years, until its destruction by the Arabs in 641 AD, Alexandria stood at the cultural and commercial center of the Hellenistic world. After the fall of Alexandria, most of its scholars migrated to Constantinople. According to one estimate, the library's entire collection of nearly 700,000 rare manuscripts was burnt to ashes."* The Islamic clarion call was, if the library books contain what is already in the Quran, then they are useless; burn them up. On the other hand, if they contain any thing contrary to the Quranic teachings, then they must be burnt any way. Hundreds of libraries and scholars have met the fate of Alexandria over centuries.

In any period of history, the victors always transplant their culture and try to disconnect the vanquished from their heritage. The reason is simple; it then becomes easier to govern the defeated people. The Spaniards burnt away thousands of records belonging to the Mayan and Inca civilizations of South America. Though the British, in their colonies, did not make huge bonfires, but they effectively used the scholars, missionaries, professional organizations, and native opportunist scholars to distort local history and heritage. It happened in India, and continues in all ages and places. Only methodology changes.

However, there are deeper considerations too. My analysis, after the Omani experience, is that if a large population of a society becomes scholarly, philosophical, theoretical, and abstractionists, then the resolve to defend national values and beliefs at every level, much less, the fire in the belly to fight, is eventually extinguished from the hearts. **Men, not willing to die for anything, are eventually overrun by foreign invaders**. Such men live for their short term survival and gains, but collective obligation is lost. Tibet is a perfect example in recent times. In history, the barbarians have also ruled and established some of the great empires. In present times, the Taliban and Al Qaeda may provide examples for future.

I recalled to Ahmed the annihilation of the educated and professionals in China under MaoTse-Tung. Mao used the youth in the Red Army to round up the dissidents and elitists, and sent them to the rural re-education camps. In China, it was called the Cultural Revolution of the 1960s. But it was a holocaust of human intellect. Subsequently, in the 1970s, Cambodia, under Pol Pot, butchered thousands of its educated citizenry. Cambodia has a museum of skulls, the only one of its kind in the world, so far.

In my opinion, the emphasis on the Quranic studies balances the so-called modern secular education, particularly of the sciences and mathematics. The growing number of *madrassas* (Islamic religious schools) in India and Pakistan, and the young men coming from all over the world to study in Pakistani and Afghani *madrassas*, until recently, indicates their popularity, and perceived meaningfulness in life.

March 30, 2009 (Oman)/July, 2014

PS 07/11/2014: A version of this Reflection has appeared in my book, *Vectors in History* – published in 2012.

A CHALLENGE MET!

It has been a month, but this incident seems like sitting on a front burner. "What do you know about differential equations (DE)?" thus said, Said Tantawy, when I asked for looking at the final exams he had graded. The University of Nizwa (UN) has a co-examiner for each course. The purpose is to ensure fairness in the grading of the final exams. Unlike at the University of Nevada Las Vegas (UNLV), the students are not shown the final exams, unless they grieve over the course grades. However, I like this idea of co-examiners for academic reasons that at least one colleague would know how another has taught a course.

Said, an Egyptian in his mid 50's, earned his PhD from an Egyptian university, though had gone to the US, perhaps, for PhD, but returned to Egypt after an MS. We shared a small office, and often chatted over the daily events. For a moment, his response irritated me, as having done PhD in Applied Mathematics, DE are my bread and butter. But gulping his remark, I laughed it away. Said was not entirely wrong either, since he only saw me teaching hard core pure mathematics courses on *Group Theory*, *Number Theory*, and *Advance Linear Algebra*.

Did I ask for these courses? No! I accepted the visiting position in July, 2008, but my spring assignment was communicated only 5-6 weeks before the classes were to begin. On asking, if there was any room for a change, the Chairman said, "Your cooperation to teach these courses will be appreciated, as students need them, and nobody else is there to teach them." Being in India at that time, I bought personal copies of the textbooks for a head start. At the end of the semester, they were given away as awards to the top students in respective courses.

I accepted this assignment to test of my teaching forte, as I always pride myself in having taught the largest number and largest variety of courses, perhaps, in the university. At UNLV, teaching assignments are often decided in consultation with the faculty members, but in the interest of the students, a faculty member may be asked to teach a specific course(s).

At Indiana University (IU), during 1968-70, three graduate courses in *Modern/Abstract Algebra* were taken using textbooks by Hernstein and Lang. I must be good at it, since the noted algebraist, Maria Wonnenberger asked me to do PhD under her guidance. The point is that if one has excelled in any activity, whether physical, intellectual, or spiritual (?), then its touch comes back quickly, even though a gap may be long. The first and the last time, I taught *Group Theory* was in Spring 1977, using Lederman's textbook. It was 32 years ago!

I took *Number Theory* from number theorists, (Late) HR Gupta and RP Bambah, during 1960-61, at Panjab University (PU), Chandigarh. I may have scored the highest marks in the University. It feels good to recall this performance even after 48 years. Chairman Gupta offered me research scholarship for PhD. Right away, I declined it in favor of IAS (for elite Indian Administrative Services). However, the last and the first time, a *Number Theory* course taught was in fall 1976, using Andrew's textbook; again 32 years ago!

Advance Linear Algebra has been taught a few times, but the last offering was 8 years ago. Lately, I have been teaching courses in DE, Calculus, Discrete Math, History of Math and Survey of Math Problems and Linear Algebra. Also, I teach Honors Seminars; *Mathematical Thinking in Liberal Arts*, *Paradoxes in Arts, Science and Mathematics*, *Non-European Roots of Mathematics.* All of them have been designed and only offered by me. These seminars emphasize history, interdisciplinary approach and surveys.

Teaching without any accountability is like shooting in darkness. I have been strong exponent of students' evaluation of the courses, and use of other instruments like grades, class averages, dropouts, contact hours etc. My mission at the UN was to have cultural experience both in academics and public. Thus, a lot of time was spent in writings and travels. A fact about mathematics is that it remains on finger tips, if done every day. I worked hard on my lecture materials, but was often only one lecture ahead of the students. The main advantage I had over them was that I could recall how I had enjoyed, wrestled and struggled with certain topics, years ago. I re-lived, and tried to convey those moments.

Nevertheless, I was never challenged by the UN students. In order to finish 135 credits (UNLV requires 124) for bachelors including 74 credits

(UNLV, 39) for math majors, they often took 18-21 credits. Being the first Omani generation of college students, they have no study skills. They nearly walked in blank to the classes next day. Initially, it was frustrating, but then I understood history and culture of the land. The Omani view on education is very different from that of Indians and Americans.

Finishing of the syllabus is not the same as perceived in places that I have taught in India, US and Malaysia. I had to tell myself and the students that *'Let us do little, but do it very well'*. On the top of students' weak academic background, the administrative enforcements are very lax. The net instruction days hardly add up to 8 weeks, not 15 weeks! Despite such boundary conditions, my teaching and students' learning did achieve their objectives.

Every good story has a point of inspiration. In 1965, I learnt that JN Kapur of Delhi University has taught every mathematics paper (equal to a two semester courses) at MA level. It struck my heart and intellect in an era (1959-61) when professors at PU could teach only one paper falling into their so called pencil-thin research expertise. This is due to a flaw in non-US doctorate programs. They may have depth for research, but lack in breadth, when called into interdisciplinary research, or teaching a random course. I do it, because at IU, I took at least two graduate courses in five of the seven areas of mathematics offered there. Also, one has to choose two minor areas besides a major. That requires additional course work. Finally, one had to pass three four-hour written exams before admission into doctoral candidacy.

I know how Said is going to feel when he reads this **Reflection**. The vast majority of the UN faculty is recruited from the countries that sell more PhDs for the Omani Rials. A paradox was observed in a sense that, whereas, the UN is a prototype of a University of Wisconsin campus, but I didn't encounter any American faculty or administrator there. Incidentally, the students are all Omani, but no Omani math PhD yet! These are some of the highlights of teaching in Oman. However, the UN is making good strides and correcting the courses, where necessary. It shall meet the current accreditation challenges, as I have met mine!

June 20, 2009/July, 2014

COMMENTS

Dear Satish, Thanks for the reflection. How are you doing? One of the words you mentioned, caught my attention, and I am writing about it. My field is Physics, but over the years, both out of necessity and 'falling in love', I am largely self-taught in math. Typically, I taught myself mat -. analysis solely for the book by Shanti Narayan. Similarly both vector algebra and vector calculus. My desire to learn general relativity has continued throughout the years, and I am now trying to learn tensor analysis and differential geometry.

I have always liked the prim and proper, and straightforward approach of our Indian textbooks. Shanti Narayan's books are excellent and I have a copy of 'almost all' his books. The text on vector calculus is by him and J.N. Kapur. I would like to know what is the latest edition, because perhaps Mr. Shanti Narayan is no more. Any comments from you would be welcome. S. Chand and Co. is a publisher. I wish they had a website to browse their stock etc. I do have friends visiting India every now and then who may oblige. Thanks and regards. **RAJA**

In this note, and also in your June 20th reflection, you make reference to unmotivated and/or unaccountable and/or incompetent profs. I've seen my share of them, too. There's a recent book out by Mark Levin, called *Liberty and Tyranny---a Conservative Manifesto*. He ends the book with ten brief recommendations about Taxes, Judges, Environment, Immigration, Foreign Policy, etc. His recommendation on GOVERNMENT SCHOOLS goes this way: "Eliminate monopoly control of government education by applying the antitrust laws to the NEA and the AFT; the monopoly is destructive of quality education and competition and is unresponsive to the taxpayers who fund it. Eliminate tenure for government school teachers and college/university professors, making them accountable for the quality of instruction they provide students. Strip the Statist agenda from curricula (such as multiculturalism and global warming) and replace it with curricula that reinforce actual education and the preservation of the civil society through its core principles. Eliminate the federal Department of Education, since education is primarily a state and local function."

How would that go over on the agenda for your next departmental meeting? **Owen**

INVERSE PROBLEMS IN
ARCHAELOGICAL MATHEMATICS

Inverse problems in differential equations distinguish applied mathematics from pure mathematics. Aside from the technicalities of its definition, they solve real-life problems. Some examples of inverse problems are: locating underground oil before digging the wells, knowing the topography of the surface of the Moon or Mars before any spacecraft lands there, locating submarines in the depth of oceans; in modern medical sciences, CT (Computed Tomography) Scan and MRI (Magnetic Resonance Imaging) Scan are other examples. The heart of inverse problems lies in getting to the source object without direct contact. Putting it in sports perspective - in American football, finding a trajectory that completes a pass from a scrambling quarterback to a running receiver is **not** an inverse problem!

Present high-tech archaeology is bringing out the richest inverse problems, when lost, 'non-existent' ancient societies, languages, or cultures are re-constructed. The scant data is painfully and expensively extracted. Archaeology, whether ground, oceanic or atmospheric, is the most comprehensive interdisciplinary area of research as well as a viable commercial enterprise. An average archaeological project engages applied mathematicians, scientists, linguists, statisticians, sociologists, anthropologists besides a legion of native laborers digging the earth, sifting it with surgical attention. Also, the students are involved in data collections, entries, and analyses under the guidance of the experts. The outcomes are reports, dissertations, and roving exhibitions.

These thoughts raced my mind when I visited the De Young Museum in San Francisco to study **Tutankhamen** exhibits, running through 2009 (known in the US as **King Tut** of ancient Egypt). It was after summer teaching, and my 5-year old grandson, a fan of King Tut, came along. The exhibits were stunning and thematically organized in different rooms. The mind would quickly lock in the re-construction of mathematics and sciences of that era that I would lose sight of my grandson. **My hypothesis is that science, mathematics and engineering flourished**

4000-4500 years ago. Yes, one has to extract any specific knowledge only out of the artifacts discovered.

An inverse problem in history gets sharper and challenging, as the time line stretches beyond a millennium. For instance, there is no human artifact that is, say, 7000 years old. The mighty Time destroys everything. It may be called the *Second Law of Thermodynamics*. Modern mass printing is only 500 years old. A couple of millennia ago, the written material, known to exist in some civilizations, was extracted from the barks of certain trees, pulps of tropical trees, roots of some weeds like papyrus (the origin of the word 'paper'), and secretion of silkworm for cloth. **Human intelligence is constant on Planet Earth**. It does not follow a linear or Darwinian model despite mutation of genes. However, there are quirky historical facts. Recently, I read that at one point in the history of Samarkand region, paper was more expensive than gold!

A few salient facts from the exhibits, though known for years, are the following: 1. The Khufu Pyramid, built 4000 years ago, is 450' high; required 2,300,000 blocks of stones each weighing 2.5 tons (5000 Lbs) 2. The human mummification took 70 days, and some mummies have lasted thousands of years. Call it amazing that intestines, liver, stomach and lungs were taken out of the body for separate preservation. 3. King Tut ascended the throne on 1336 BC at the age of 9 and died at 19. Howard Carter, the British Egyptologist discovered Tut's tomb in 1922. His coffin made of solid gold weighs one ton (2500 Lbs)! During that era, all the gold belonged to the pharaohs, unless it was gifted by them. Tut's throne is made of gold, inlaid with ivory, precious gems and glass. **It matches with any piece of craftsmanship in the world today!** The monuments and such artifacts are the DNAs of math and science of that era.

Here is a viewing angle. Take for example, the present greatness of American society. It is mainly due to the fact that the US tops the world in math, science and technology. **They go hand in hand for a while, since the rise and fall of civilizations is a lesson of history.** It is reasonable to assume that science and math existed in a society that had the knowledge to build incomparable pyramids, mine gold and minerals, cut precious metals and boulders with the highest precision etc. Their applications may be different. An absence of archaeological evidence of modern tools or techniques implies that the 'undiscovered ancient tools and techniques' were no less great. A few years ago, a multimillion dollar project of the National Geography Society of building a 24'-high pyramid using assumed BC construction techniques, failed badly!

For decades, a theory of free slave labor was propagated as a simplistic explanation of the construction of the pyramids. It has now been completely debunked, but it may take a few years before it is excluded in new books and journals. Partly, it was a colonial conspiracy to depict the colonized nations as inferior. A conjugate theory is that **the natives are not real natives**, and hence the colonizers are not the occupiers of the lands! As a corollary, sometimes the colonizers called themselves emancipators, and the colonized ones, like white man's burden. The entire decolonization of Asia, America and Africa has destroyed this myth too.

One of my recurring thoughts, while watching or reading about a great ancient civilization, runs like this. All of us have lost an item never to be found again; a family loses it too, or a family itself is lost, in time; an entire society and nation have been lost, and so are civilizations, in the great march of Time. Individual memory and collective memory, whether of a family, or nation have the same retentive characteristics, but different scales of time. Knowledge is also lost along with its keys. **No research is absolutely new; it is all very much a case of re-searching!**

The assumption of multiple solutions and multiple approaches to any problem sits at the heart of resolving a knotty question. However, it is a cultivated trait. A seeker of truth has to shed any perceived superiority of one solution/approach over the other. On a personal note, I encourage and reward students for doing a problem in more than just one way. A class discussion of a problem is not over after finding a solution, but exploring other possibilities.

For the last couple of years, my mind has been engaged with a question of finding the necessary and sufficient conditions for the development of science and mathematics in terms of economic, social, political, and most importantly, religious variables. It is an open question. Finally, it is worth knowing that, at present, there are places on the face of the earth that are still living in dark ages of math and sciences. **Dark ages do not exist in past alone**. Some ancient places and people were as fully developed in brain and brawn, as are some places and people today!

Aug 31, 2009

COMMENTS

Thanks, Satish. I appreciate your work / writing. Have a great semester. Be in touch. Sincerely, **Doy Hollman**, Lipscomb University

An excellent & brilliant presentation. Thank you for sparking our imagination!!!! A nice Diwali fireworks. Thank you for sharing your thought with us==**Soori**, Hoosier friend!

Dear Satish Bhai Sahib - As is the case with all your other articles, this is one of the most brilliant analyses of the past and something which sets the thought process delve into the imponderables of the past. The human mind has not till now fathomed the depth, knowledge and the scientific advancement of the past and you are absolutely right that man tends to re-search most of the time. The Inverse problems applied through differential equations linking the applied math's to a host of other economic, socio-cultural, historical, architectural issues past and present could open the doors of exciting new propositions. This is something which a trained and analytical mind like yours could excel in, and quite the opposite with a person like me with quite a narrow specialization in a small off shoot of the commercial world. **Lalit**

Hi Satish, Very well written and an interesting philosophical perspective. I agree with you – ancient people were every bit as intelligent as modern ones. In fact, I think that modern technology has made us all soft and, at least in practice, humans are dumber than we've ever been! Our world is so automated, why do we need to use our brains? Few people really have to think any more. That's why robots are slowly replacing us. Yours, **Ed**. Dr. Edwin Barnhart, Director, Maya Exploration Center 7301 Ranch Road 620 N, Suite 155 #284 Austin, TX 78726.

Great observation and I agree with the conclusion. Intellectually the generations gone past long ago were as intelligent as present one is. However as the time passes the collective human experience makes the next generation smarter. The difference is this collective human experience in both failure and success which enables the future generation to build on it. **Rahul**

Dear Professor Bhatnagar, Thank you for your kind comments on my article and for sending me your own ruminations, which I read with interest. Best wishes, **S. Prange**

ICM: A MATHEMATICANS' SHOW

Three weeks ago, I received copies of *Reflexions*, the daily newsletter of the **International Congress of Mathematicians** (ICM), held in India - Aug 19 through 27. A US friend, attending the ICM, had brought them along. Only some old timers still like to hold on to the hard copies and snail mail. The ICM website, www.icm2010.org.in, has a whole lot of updated information on papers and posters presented, people, paparazzi, prizes, places, and plethora of pictures.

Once I started reading the first issue, I could not put them away until they were all done. I enjoy extracting the humane sides of mathematics and mathematicians, considered universally dry, and connect them with national politics, religions, life styles, etc. It is my forte. When I heard about a recent book, *Mathematics and Religion* by Javier Leach, my mathematical flights and forays into remote regions of human intellect were reinforced.

The ICM is being held every four years in its 100-year old chequered history. Initially, it was mainly European to counter the JMM, the annual **Joint Mathematics Meetings** of the AMS, MAA, and SIAM. Despite its international focus, the ICM-2010 drew only 3000 delegates, which is nearly 50% of the registrants at the JMM- 2010, held last January in San Francisco. Apart from thousands of research papers, talks, and displays, the Employment Section of the JMM draws people from all over the world – irrespective as to whether the jobs are sought in the US or anywhere in the world. One can identify any ethnic face and language. I just sit, sip, and reflectively see people go by.

Based upon the names and institutional affiliations, two out of four Fields Medalists, seem to be Jews, namely, Linderstrauss and Sminiov. My second guess is that out of three other notable mathematical award winners - Spielman (Nevanlinna Prize), Meyer (Gauss Prize), and Nirenberg (Chern Medal, the newest), at least one is a Jew. **How can you not notice the statistical domination of the Jews, and stark absence of other religions?** Also, it is obvious that some political systems are more conducive for the cutting edge mathematical researches. Princeton,

Caltech, MIT, Harvard and Stanford dominate the entire world of mathematics. They epitomize American system of education at large, as did the Cambridge and Oxford universities for the British before WW II.

While, one may credit the development of great US universities to the capitalistic system, but on a small scale, the exponential growth of mathematics in South Korea, is simply unbelievable. It has already hosted 1998-Summer Olympics, and is selected to host both the ICME-2012 and ICM-2014. The combination of the US-promoted free enterprise and 'despotic' presidential rules since the 1950 'division' of Korean peninsula has made South Korea, the 10th leading economy in the world.

On a mathematical front, the young Korean Mathematical Society boasts a membership of nearly 3000 including 1500 mathematics professors. In 2008, South Korea was ranked 11th in terms of research publications. In International Mathematical Olympiads, Korean kids are ranked third or fourth. It is remarkable for a country one third of the size my US home-state of Nevada, and one half of my Indian home-state of Punjab. Its population is only 50 million.

How do you explain this all round growth except by the organizational roots of the Koreans in Buddhism and 'discipline' enforced by the political system prevailing for the last 60 years? A lesson: democracy is not the only political system that is good for all the people all the time! In comparison, look at India, the largest democracy in the world with 1.7 billion people. It has nearly 7000 four-year colleges and universities, but the annual meeting of the 110-year old Indian Mathematical Society has never drawn even 200 persons. How do you explain it except by the ultimate individual freedom, enshrined in Hindu religion? It is reflected in the un-organization of Hindu institutions at every level – from academic to political? The fact that India's population is 80 % Hindus has a far reaching impact.

Amongst the tidbits, it was amusing to learn about James Harris Simons, a non-trivial mathematician and a close friend and collaborator of Shiing-Shen Chern. Simons has made a fortune through his **Renaissance Technologies** company. With his present worth at nearly 10 billion $$, he is described as "the world's smartest billionaire." Simons reminds me of one of the pathological counterexamples – like, a function continuous everywhere, but differentiable nowhere. Personally, while growing up in poor socialistic India of the 1950s, we never dreamed of striking rich. Moreover, money was considered antithetical for academics!

Since one of the aims of the ICM is to foster mathematics in developing countries, the host country becomes a natural focal point. The profiles and interviews with superstar Indian mathematicians - Varadhan of Courant Institute - 2007 Abel Prize winner, Kannan Soundrajan of Stanford, and Kiran Kedlaya of MIT were showcased. On the one hand, it underscores the strength of the US education system, and on the other hand, it reinforces a correlation between Hindu belief system and mathematical thinking. **The freedom to choose co-ordinates in the solution of a math problem or axioms in the development of a mathematical system is akin to choosing your own beliefs in life.**

There were a few prominent obituary notices in one issue. I had heard of legendary US statistician, David Blackwell (1919-2010), but did not know that he was the first Afro-American professor to be tenured at UC Berkeley. The son of a railroad worker, he overcame every discrimination.

Amongst various sidebars, peppering the issues, IM Gelfand (1913-2009) is aptly quoted. "Mathematics is a way of thinking in everyday life. It is important not to separate mathematics from life. You can explain fractions even to heavy drinkers. If you ask them, "Which is larger, 2/3 or 3/5?'. It is likely they will not know. But if you ask, 'Which is better, two bottles of vodka for three people or three bottles of vodka for five people? They will answer you immediately. They will say two for three, of course." In the present times of fiscal crisis, selling any thing is difficult- hard mathematics is the hardest to sell! Some mathematicians have to step to the plate, as it is a question of survival.

Dec 09, 2010

MATHEMATICIANS IN MANHATTAN PROJECT??

"Physics won WW II," is a claim commonly heard ever since I got interested in reading about it. Early on, during high school and college years in Bathinda (a small town in Punjab), my heart was in physics for weird family reasons. Due to limited curriculum, I ended up studying mathematical physics for my bachelor's and master's in India. While attending a Chautauqua course (Oct 04-06), *The Birthplace and Early History of the Atom Bomb*, I heard the same yarn on the 'supremacy' of physics.

To some extent, this claim does make sense. The most famous mathematics equation, $e = mc^2$, as derived by Albert Einstein, is the heart and soul of any nuclear bomb. However, in the shadows of great physicists who were engaged in Manhattan Project, there were top-notch chemists, engineers, metallurgists, and no less, the US steel factories delivering incredible pieces for the 'gadget', as the atom bomb was secretly referred to. Manhattan Project was the biggest symphony of America. My curiosity is as to who played mathematics in this orchestra!

Also, the spotlight had to be on physics as, Robert Oppenheimer, the scientific director of Manhattan Project, was a well-known physicist. The Bradley Museum in Los Alamos displays the pictures of 60-70 pivotal men and women from every phase of the Project. I don't remember any one identified as mathematician, unless my roving eyes missed them on the honor wall. Of course, I was thinking of mathematicians of the Fields Medal stature. Like most people, I was not aware that the Project lasted only 27 months, as it was officially closed in September 1945. However, the next phase, on the development of Hydrogen Bomb, continued in Los Alamos and in other associated national labs.

A couple of days ago, while reading *Rider of the Pale Horse* (2005), I noted that mathematician, Stan Ulam (1909-1984), a Polish Jew, had the highest security clearance in Los Alamos lab. The book, distributed as a part of course materials, is written by McAllister Hull. Drafted in the Army at age 20, Hull worked in Los Alamos in several engineering phases of the atom bomb. After the war, he earned his PhD in nuclear physics, taught in Columbia University, and went on to serve as Provost for 23 years at the University of New Mexico, Albuquerque. He wrote this book at the age of 82 with the assistance of a writer and his artist son. The book is captivating, as it combines the technical nuts and bolts with a few human dramas. Hull was the only 22-year old non-VIP to witness the first detonation of the atom bomb on July 16, 1945.

Mathematicians were hired in the Theoretical Division of the Los Alamos lab. Most were young Jews in their 20s who had fled Europe. On landing in the US, they would, like most immigrants, seek the advice from fellow compatriots. Invariably, they were keen on assisting the US in the war efforts, and hence they found their way to Los Alamos, which did not exist on any US map then. I think J. Carson Mark was the first mathematician who headed the theoretical division.

A few other mathematicians that I could search out are John Hinton, David Frisch, and Joseph McKibben. Of course, the legendary mathematician John von Newman (1903-1957), a Hungarian Jew and one of the four founding professors at the Institute of Advanced Studies at Princeton, was actively engaged in the work at Los Alamos. Besides, he directed many incoming European mathematicians towards Los Alamos.

Two things stand out in my mind. One - nearly a dozen physicists, who researched in the labs of Los Alamos, eventually won Nobel prizes in physics - based on their researches and discoveries in nuclear fission and fusion processes done in the development of atom and hydrogen bombs. I am sure not all researches were declassified right after the war was over. However, I don't think any mathematician won a Fields Medal. This may have been partly due to the fact that the Fields Medal was instituted in 1936 and is not awarded to mathematicians over the age of 40. Also, no awards were given out between 1936 and 1950.

The other thing that flashed my mind was the corresponding role of the British mathematicians in the War. Cambridge University was the most celebrated place in UK and GH Hardy was the doyen amongst mathematicians of that era. Years ago, I remember having read that Hardy had walked out of a meeting of the army generals with scientists and mathematicians. It would be interesting to know the circumstances surrounding this incident.

With a few exceptions, the US universities were hardly known internationally even after the WW I. During WW II, the collaboration of the scientists and mathematicians in the war efforts created new ties and synergy. After the War, the US Army, Air Force, and Navy routinely funded the universities for doing researches on both pure and applied mathematics. This collaboration went on through the 1960s. But it received a big blow when the antiwar (Vietnam) protesters occupied the university campuses. Later on, the Middle East oil embargo of the 1970s brought the US economy to nearly its knees for a while.

On a personal note, my doctoral research at Indiana University was done under an Army research grant awarded to Prof Robert P. Gilbert, my thesis supervisor. It seems, once in lifetime, a Manhattan Project is essential in the life of an individual as well as that of a nation!

Oct 13, 2012

COMMENTS

Stan Ulam is a fascinating man. He was willing to work on Super. Von Neumann helped to bring Ulam to Los Alamos. Ulam died in Santa Fe. The Vietnam War did not totally end the Army's support for mathematicians. My close friend continued to get funding from the Army. There was a shift away from universities to places like the National Bureau of Standards. Explain further about the impact of the oil embargo. It hurt, but I think you exaggerate when you say it brought the economy to its knees. Yes, it created world-wide inflation, very bad in the US. Serious consequences. But it was on top of the cost of the Vietnam War. **Noel Pugach**

2. The main bit of reading that I've done on the Manhattan Project is *Genius*, by James Gleick. It's a biography of Michael Feynman, a physicist about whom Oppenheimer said that he'd sooner lose any two of the people working under him than Feynman. The scientists and mathematicians at Los Alamos were divided into teams, and I found it interesting that the team Feynman worked on was studying the obscure sounding topic of diffusion.

I'm going to check out the book you mention, Rider of the Pale Horse. I did a Google search, and read through nearly a dozen references to the mythological four horsemen of the apocalypse before I got to Hull's book. Regarding the physicist/mathematician split, my opinion is that no one can be a physicist without really strong mathematical ability, but a mathematician can make great contributions in the field without doing work that is directly related to physics. My suspicion is that most of the mathematicians who worked at Los Alamos were chosen because their area of specialization had something to do with theoretical physics.

You mention that your doctoral program was done under an Army research grant. I got my Master's at Wisconsin as a participant in an NSF-funded program called the Academic Year Institute for Outstanding Science and Math Teachers. Best wishes, **Owen Nelson**

SECTION V

OTHER PERSPECTIVES

WHAT HOM MEANS TO YOU!
(A Reflective Note to Colleagues and Mathematical Friends)

This is a kind of appeal for your contribution on History of Mathematics (HOM) – irrespective of whatever your perception is of it. In case, you are wondering where this call is coming from. Then let me tell you that I have been working on a book on HOM for the last four months. It is unlike any other book on the subject – far from being a typical textbook on HOM. On the other hand, it has discrete and diverse reflections on every conceivable angle on HOM.

It is appropriately titled as ***Darts on History of Mathematics***. It is an evolutionary offshoot of my interest in HOM spanning decades, and viewed as a subset of history in general. Of course, I have designed and taught courses on HOM - both at the graduate and undergraduate levels.

An underlying fact of life is that interest in some aspect of history in general is integral to the making of an intellectual mind. Thus, a sense of history is unavoidable, as we age. It eventually rubs into our social and political issues, and in particular, on our discipline of mathematics too. I have seen it happening to most mathematicians in their grey years. As a matter of fact, I have noticed very distinguished mathematicians of the caliber of Abel prize winners sitting and actively participating in the sessions of HOM at the annual joint math meetings held in Jan.

There are no boundary conditions on this write-up except it be in at least 200 words – no upper bound! In order to avoid multiplication, they may be gently edited. Other than that, if you have any personal encounter with HOM, it is most welcome. History of any mathematical topics of your fascination and research are great. On the humanistic side, included are the lives of mathematicians, math teachers who may have touched and inspired you. It could be from your experience of having read or taught HOM – not excluding are the curriculum trends and teaching etc.

The book is in the final stages and this idea just flashed my mind. Often, the books have quotes and excerpts from other books of established authors, and they are stacked at the end of chapters and in a bibliography at the end. It is far more than padding the material. In contrast, I decided to have them only from my colleagues and friends who also have spent a good time in mathematics. That will make the material more vested and vetted.

Is there a dead line? Yes, within two weeks - while this request remains fresh. In our tweet and texting age, if we don't get on it sooner, it will just fade away from the mental screen – like, hundreds of running ads seen on the PCs and on all kinds of billboards. So, get on it right away

Thanks in advance!!

Aug 03, 2104

ORTHOGONALITY OF MATHEMATICS & HISTORY

Using a scenario of kids' upbringing, mathematics and history are born to be apart - raised differently in a sense that for the first 15 years of life, math is treated like a privileged one, as child prodigies are found mostly in math, besides in music. History, for all intents and purposes, is not even born then. In high schools, math feeds on deductive thinking and problem solving. Whereas, history bogs down the students in memorization of the inane dates of the life spans of history makers and events.

Different types of mindsets are required for each discipline. The intellectual training diverges in college. In the US curriculum, a math major does not take a history course beyond one or two required ones in the general education core. Likewise, a history major rarely takes a course beyond a rudimentary math course - again required in the general core. This fork widens up when they go for graduate degrees.

However, a funny thing happens on the way to adulthood. While growing up in life, one cannot absorb math from the air. Math demands serious commitment in studying. For example, for freshmen students, two hours of studying for every class hour is a must. On the other hand, history is associated with a mature mind. A saying goes: history is for the wise. That is why it is ludicrous to talk of child prodigies in history. Nevertheless, every adult absorbs history by reading books and magazines, debating and listening to talk show hosts on radio and TV. Essentially, you don't have to have a history degree to understand and participate in any aspect of history. However, math defies and frustrates any self-taught learners.

This preliminary analysis of the nature of math and history struck my mind 4-6 weeks ago when I received only five open ended write-ups on the history of mathematics. An email was sent out to nearly 300 math professionals - starting from math graduate students, teaching assistants, part time instructors, colleagues, friends and acquaintances in various positions in India and the USA. Some are world renowned mathematicians, department chairs, chair professors, former presidents and officers of Indian and American mathematics societies. Each one is personally known to me. I did not want to include or quote any stranger irrespective of his/her reputation. In many cases, it was followed by another e-mail or phone calls. There was a promise, but no delivery. That is baffling. It is known that creative writing is not mathematicians' cup of tea - thus are people often surprised at my writings.

It may be interesting to add that amongst 30+ of math faculty at UNLV, five are from China, three from India, two each from Iran, Sri Lanka, Russia, Taiwan; one each from Brazil, Canada, Korea, Haiti, Mongolia, Bulgaria, Serbia. Besides, personal notes to most of them, I urged them to write on any history of mathematical topic preferably related to their national origins in order to spice up the material. It did not generate even a single response or acknowledgement in writing or in passing in the hallways. The competitive US academic environment breeds weird collegiality and collaboration.

This refection is a small attempt to understand the non-intersecting nature of the curves describing math and history. GH Hardy (1877-1947) was the first mathematician who put it in black and white that creativity (Fields Medal type) in math starts sliding down after the age of 30. I liken such research mathematicians with heavyweight boxers. That is why, mathematicians over the age of 40 are not even eligible for the Fields Medals. Ironically, this is the age of the coming of budding historians!

Because of the conditioned mind and institutional pressures, research mathematicians continue to spend their professional lives on a beaten path and hope for a 'Viagra' moment. At the annual JMMs, I have seen mathematicians of the caliber of Abel Prizes sitting in on the sessions of history of mathematics. It is in the lower tiers of institutions, where math professionals branch out into other areas once they are free from the pressures of getting tenure and promotions. This is amply supported by faculty participation in Chautauqua courses from very small colleges and universities. Most of the time, I am the only one from a PhD granting institution. Fortunately, my learning curves of history and math have many points of intersections due to my bizarre path of education. In general, there is no blue print or template for the making of a historian of mathematics.

I venture to remark, that after 50 years of age, mathematicians are psychologically ready to branch out – be it in history of mathematics, mathematics education, or philosophy of mathematics. Unfortunately, in the US, there is no joint curriculum in history of mathematics for degrees. Interestingly enough, some UK universities do have it.

It is only the tip of an iceberg - whether considered as research problem or an educational tidbit. The five writers include one emeritus math professor, one physics professor and author of history of science books; two are my students from history of math course taught a year ago, and the fifth one is a PTI. Their write-ups are going to be included after some modifications.

Sep 12, 2014.

[Mangho Ahuja, 79 years old, is an emeritus professor of mathematics from Southeast Missouri State University in Cape Girardeau, Missouri. Our acquaintance goes back to 20-some years. It primarily rests on our common Punjabi upbringing and education received from Panjab University, Chandigarh. We often meet at the annual Joint Mathematics Meetings, where we would exchange tidbits of life in general.

Mangho's write-up is the only one received from a mathematics professor. At the same time, his write–up briefly explains why it is not easy for mathematicians to write on any topic of history of mathematics. In addition, any writing skill that survives through bachelor's degree is muzzled out during graduate studies of mathematics, as mathematical writing gets linguistically cryptic and compact. Eventually, one is distanced from creative writing. To make it worse, the demands of publish and perish for merit, sabbaticals, tenure and promotion do not leave any time and mood to study history of mathematics independently.]

BITS AND BYTES ON HISTORY OF MATHEMATICS

Thanks for your telephone message. You had called about a piece of 200 words or so, on the History of Hindu Mathematics. You said: write whatever you know, whatever comes to your mind. I don't mind the work, but the assignment is a little too vague, and I am not sure what I can write on. Let me try to be a little more clear.

Yes, I am a fan of History of Mathematics. Yes, I know something about Hindu mathematics. No, I did not particularly study Hindu mathematics. For one of my projects, I had to study Islamic math. Most of what I know about Hindu Mathematics is found in every ordinary book on the History of Mathematics. Anything I write will be a copy of what the books say. I admire you, Satish Sahib, for the new thoughts you think of and new ideas you discuss in your *Reflections*.

Let me add few thoughts, or a personal history of my own. This is certainly not the assignment you have asked for. Sorry. But I am sharing with you, as a friend, my personal history. I really do not know how to classify what I am writing below.

I passed BA in Sept. 1954, and MA (Math under old regulations) in Sept. 1958 - both from Panjab University. For neither of these exams, was I supposed to read/learn History of Math. It was not required. I spent 3 years (1963-66) at the University of Rochester, Rochester, NY, where I got another Master's Degree, and finally got my PhD from University of Colorado, Boulder in 1971. Nowhere was I required to learn/study History of Mathematics.

In 1969, a professor friend at Boulder was teaching History of Mathematics. He came to me one day and said: Mangho, you are a Hindu, aren't you? I said: Yes. He said: You will be giving a lecture to my class on Hindu mathematics. I said: OK. I rushed to the library and got a few books on Hindu Mathematics. Among others I remember the book by Dutta and Singh. This is the first time I actually read a book on Hindu Mathematics. We all know about the contribution of Hindus – the Zero, the place value system, and few other things.

For my Ph. D exam, I never had to learn History of Mathematics. But in the Oral exam (defense of dissertation), one professor said: I am going to ask you some questions on the History of Mathematics. One day, I am going to flunk the student (not you, Mangho), who does not answer these questions correctly. Well, the questions were easy. The point of all this is that, in my case, History of Mathematics, as a subject was totally missing from my mathematics curriculum right up to Ph.D.

In 1968 with ABD (All But Dissertation), I joined Southeast Missouri State University. I learned that to be a High School math teacher, History of Mathematics was a required course. No High School teacher in Missouri can be certified without completing this course. Around my retirement time in 2002, I taught this course a few times, but I was not very successful.

In 1990, I was working on a project on Islamic Calligraphy. For this, I read a lot about Islamic mathematics, culture, arts, etc. Again, my interests deviated from Hindu Mathematics.

Today, people know more about Ramanujan and his Notebooks. Thanks, mainly to the British and American mathematicians who wrote fondly about him, even though they are still trying to figure out how Ramanujan got the results that he got.

While most people in the world would call me an educated person, I feel so humble and so ashamed that I know so little mathematics, History of Mathematics, or History of Hindu Mathematics. I feel ashamed because people "expect" me to know it since I am a Hindu, and I am a mathematician. My only defense is that neither the educational curriculum, nor my own career interests in mathematics, required me to know the history of Hindu Mathematics. So, I remain ignorant till this day. That is all.

[Alok Kumar, age 60+, has been a professor of physics at the State University of New York, Oswego since 1992. He has served for six years as Department Chairman, which is uncommon amongst the Hindu professors. Besides, a prolific physics educator and researcher, he is an established science writer for having published two scholarly books and numerous articles on history of science. The first book deals largely with science in Middle East, and the second one focusing on ancient India is just released. We have known each other from our mutual writings for over 30 years, though have met only once. However, our exchanges of ideas on phone are long and regular. Alok is one of the rare Indian academicians in the US who have made their professional marks with Indian PhDs. He did PhD from Kanpur University in 1980.]

TRANSMISSION OF HINDU
MATHEMATICS TO MIDDLE EAST

The exchange of knowledge between India (known in medieval times, as Hind or Hindustan - meaning the land of the Hindus), has spanned over a millennium. This migration of knowledge has played an important role in the growth of science, particularly in the European Renaissance. While the scholars have studied Greek encounters with other cultures in great details, the exchanges between India and Middle East have not received adequate attention.

Al-Khwārizmī (780-850), Al-Uqlīdisī (920-980) Ibn Labbān (971-1029), Al-Bīrūnī (973-1051), Ibn Sīnā (980-1037), and are some of the leading Islamic scholars of the medieval period in Middle East. These scholars are known for preserving as well as transmitting mathematical knowledge of the Hindus to Islamic Middle East. Above all, they were quite truthful in documenting Islamic knowledge that was based on Hindu mathematics.

Popularly known Hindu-Arabic numerals were first created by the Hindus, as the name suggests. They were adopted in Arabia, and later on, transmitted to Europe. Other cultures struggled with mathematical calculations because of their inferior numeral systems. Today, the Hindu numerals are universally adopted. It is for this reason that Copernicus, while suggesting a heliocentric system for the solar system, used the Hindu-Arabic numerals since it allowed fast calculations.

One of the earliest written records of the place-value notation comes from Vāsumitra, a leading scholar in Kanishka's Great Council (ruled 127–151 AD). Five centuries later, it was corroborated by Xuan Zang (also known as Hiuen Tsang, 602 - 664), a Chinese scholar who travelled across India with royal Chinese patronage. He wrote that king Kanishka convened a convocation of intellectuals for a scholarly compilation, which is known as *Mahāvibhāsa.*

Hindus, with their numeral system performed mathematical calculations that eventually became a mark of modern science - including in the determination of the age of the universe, the total number of species in the universe, the total number of atoms in the universe, etc. Since other cultures had no such perception of large numbers, Al-Bīrūnī, after visiting India, wrote in his classic book, *Alberuni's India,* "I have studied the names of the orders of the numbers in various languages with all kinds of people with whom I have been in contact, and have found that no nation goes beyond a thousand. The Arabs, too, stop with a thousand, which is the most correct and natural thing to do. Those, however, who go beyond the thousand in their numeral system are the Hindus, at least in their arithmetical technical terms." However, with the evolution of modern science, the use of large numbers has become a common place. The ancient Hindus were simply ahead of their contemporary cultures, as the US society is today.

Once the Hindu number system was introduced to other cultures, they had to compete with the prevalent local numeral systems. This happened in the Middle East and Rome. Some nationalists did not like this foreign system as it was indicative of Hindu intellectual superiority. Thus, the decrees were issued against them, loyalties were defined by their users, and lobbies were formed where the superiority of one system over the other was debated.

The reason for not accepting Hindu mathematics was simple. Around the 11th century, while Europe was beginning to come out of dark ages, the sun was beginning to set on Hindu civilization. Most Hindu kings in the west and north of India had lost their kingdoms to the Muslim invaders from Middle East. After all, who likes to follow any traditions of the vanquished?

For instance, Severus Sebokht (575- 667 AD), a Syrian orthodox Christian Bishop and Aristotelian scholar mentions of such a rivalry between the Greek and Hindu numerals, "I will omit all discussion of the science of the Hindus, a people not the same as Syrians, their subtle discoveries in the science of astronomy, discoveries which are more ingenious than those of the Greeks and the Babylonians; their valuable method of calculation; their computing that surpasses description. I wish only to say that this computation is done by means of nine signs. If those who believe, because they speak Greek, that they have reached the limits of science, should know these things, they would be convinced that there are also others (Hindu) who know something."

Since Prophet Muhammad, the founder of Islam, lived from 570-632 AD, it is conclusive that the Hindu numeral system was already known in Arabia before the advent of Islam. The Arabs had discarded the Greek numerals in favor of the Hindu numerals.

Another instance is of Ibrahim al-Uqlīdisī (920-980 AD), also written as Uklidisi, means "The Euclid-man" for his role as a copyist of Euclid's work. His book, *The Arithmetic of Al-Uqlidisi: The Story of Hindu-Arabic Arithmetic as told in Kitab al-Fusul fi al-Hisab al-Hindi*, explains arithmetic through Hindu numeral system: "I have looked into the works of the past arithmeticians versed in the arithmetic of the Hindus . . . We can thus dispense with other works of arithmetic....." The title of the book indicates that it contains mathematics of the Hindus.

Talking of the books, Al-Khwārizmī also wrote a book in the Arabic on the use of Hindu numerals, called, *Kitāb al-hisāb al-Hindī*. The book is now lost, however, its Latin translation by Robert of Chester has survived. It is a compilation of the mathematical methods used by the Hindus. He acknowledges; "The Hindu method of calculation (*hisāb*) with nine digits. With these digits arithmetic, *i.e.* multiplication, division, addition and subtraction, may be simplified." Al- Khwārizmī was the most influential Islamic scholar before European Renaissance.

Furthermore, Kushyar Ibn Labbān wrote a popular book on Hindu arithmetic, *Principles of Hindu Reckoning*. Interestingly, he started this book with the following sentence: "In the name of Allāh, the Merciful and Compassionate. This book on the principles of Hindu arithmetic is an arrangement of Abū al-Hasan Kushyār ibn Labbān al-Jīlī, may God have mercy on him."

Finally, the concept of emptiness is highly valued in Hindu and Buddhist philosophical thoughts, and is defined by the Sanskrit terms *sunyata* and *neti-neti*. In Hindu philosophy, God is defined as someone with no attributes. This is where the concept of *neti-neti* (not this, not that) comes from. This "emptiness" evolved into a mathematical reality in the form of zero. When the concept of zero migrated to the Middle East, Arabs literally translated the Sanskrit term for zero, *sunya* (empty) into the corresponding Arabic word *sifr* (empty). This Arabic term became cipher in Europe. Leonardo Fibonacci called it *zephir* in his book, *Liber Abaci*. The word has since evolved into zero in English.

[Adams Koebke, age 31, is a fully home-grown mathematics student in a sense that he was born and raised in Las Vegas, and he received BS in mathematics from UNLV in 2011. He took two graduate courses from me - including one on *History of Mathematics* (MAT 714). It is required in the **Teaching Concentration** - one of the four concentrations in the MS program. He expects to finish it by December, 2014. Afterwards, his plan is "to continue to serve math community by helping undergraduate students complete their math requirements, whether directly in the classroom setting or indirectly in the form of developing learning centers and remediation resources."]

HOM IS WHERE THE HEART IS

Throughout my education, mathematics has always been my strongest subject, but I can remember becoming enthralled with the drama and significance of history at an early age. It wasn't until early in my graduate studies when I began teaching that I ever thought to look deeply into the intersection of the two sets. And, what I found is that they complement each other beautifully. The common casts of characters in HOM (History of Mathematics) have storylines full of plot twists and rivalry, success, failure, and redemption.

Examining the growth and development of a branch of mathematics over time allows us to better appreciate its modern state. To forget about those stories and people is undermining its significance. I believe, as an educator, it is my responsibility to share HOM with students, integrating these stories within the curriculum, so they see that this beautiful science wasn't built in a day, and hasn't stopped evolving. To hold in high regard the great minds of the past inspire students of the present to strive to leave a legacy behind, just as those that came before them have done.

[Shaodong Lin, age 30, is a foreign student from China – like, I was when I came from India to Indiana University, Bloomington in 1968. In 2008, after graduating from Dalian University of Technology, Dalian, he came to the US and attended University of Minnesota and South Dakota State University for graduate studies. In 2012, he transferred to UNLV, and took my course on *History of Mathematics* (MAT 714. After master's in Teaching Concentration, he plans on going for PhD.

Shaodong's following write-up is the result after editorial touches upon its grammar and syntax, as his command over English language is still in progress. Inclusion of this simple piece is likely to encourage him to do more research]

REFLECTING ON HISTORY OF MATHEMATICS

History is the key to the success; people learn and master the skills and knowledge from their ancestors. I'm lucky to have chance to take a course on history of mathematics and see its importance as to how it influences modern mathematics and mathematicians.

Learning history of mathematics is helpful for using proper methods to solve current problems. We can see how our ancestors solved some these problems when they had them, which can broad our thoughts and offer efficient methods.

Secondly, every mathematics crisis is in the evolution. This brings methods of mathematical analysis, and how to change mathematics views even for the world outlook. For example, Greek mathematician, Hippasus (5th century BC) found the irrational number exists in the diagonal of a square.

Thirdly, history of mathematics tells us that mathematics is the key to the science. There are many great physics achievements now, but there is none of them which does not rely on mathematics. Great physicist Roentgen discovered the X-ray in 1901, so he became the first winner of Nobel Prize in physics. When he was asked that how he won, he told people the secret, the first thing is mathematics, the second thing is mathematics, the third thing is mathematics.

History of mathematics is the source of success. It's one of the most interesting subjects for mathematicians to know how great people succeed.

[Owen Nelson, 82 year old, has been a part time instructor at UNLV for over 15 years – hired for the sole purpose of teaching remedial math courses. He got his MS in science education from UW- Madison and ABD (All But Dissertation) from Indiana University. Owen is one of the rare individuals whose intellectual growth has not been limited by formal college degrees. He has commented on most of my mathematical reflections. In a way, he has lived through historical events on many fronts. Often, his feedback brings out clarity in my thoughts. Owen stands out in contrast with many PhDs who seldom see, smell, talk or think out of a box of their narrow specializations.]

FUSION OF HISTORY: SCIENCE AND MATH

My field of specialization is science education, which is math-related, but not math per se. The most advanced math course that I've taught is 124, and I'm simply not qualified for anything beyond that. I had a calculus course 60 years ago, but it's a dim memory.

I fully agree that "...interest in some aspect of history in general is integral to the making of an intellectual mind." I wasn't much turned on by such courses as American History and World History taught with more emphasis on dates and wars than on the personalities of participants, but my master's program at UW-Madison included a one-credit course in the history of chemistry, taught by a senior faculty member.

You can't picture in your mind the time of the alchemists without considering the historical factors existing at the time---mostly monarchies for governments, religions and superstitions prominent (dominant?) in most societies, widespread illiteracy, agriculture - the dominant industry, little travel for most, etc. A study of history requires you to conceive of events in the world, as it was then---to put them in a context that we are quite unfamiliar with.

An interesting example. We live in a world where we not only hear about distant events, but we can often see them on TV as they happen. I saw the first man step on the moon. I heard about the first plane hitting the World Trade Center, but I saw the second plane hit the second tower. On the other hand, there were many centuries where most people were born, lived out their lives, and died without traveling more than a few miles in any direction.

There is an afterthought. I'm concerned that, since the advent of television some fifty or sixty years ago, there has been a constant decline in the activity of reading from the printed pages. In today's world, AADDDs (Addictive, Attention-Demanding Digital Devices) only exacerbate the decline. A great many people consider themselves informed because they watch TV or see something on a Facebook page or go to a movie. TV news will seldom go deeply into any topic, but it's easy for the viewer to think, "Well, now I understand that."

Giving another touch of history, I can clearly recall a discussion about nursery rhymes and fairy tales that took place in a "children's literature" class at Indiana University, probably in 1969. I doubt that any other than the professor knew prior to the discussion that Walt Disney's early animations---Cinderella for example---were severely criticized on the grounds that they took away the element of imagination. The same argument was made about illustrated versions of the tales in print form. When hearing or reading about the pumpkin coach, only the child's imagination can create a conception of it for the child, but imagination is stifled when an artist's illustration is shown. Understandably, the thought will automatically be, "That's what it looked like," and imagination is unnecessary.

Back to HOM (History of Mathematics) - I've never had a course in HOM, but I've read quite a bit. I've included a bit of it in my classes, but it's seldom (never?) included in the course objectives, so bringing it in means diverting from the main plan. My understanding is that few take it other than math majors. I'd like to see everyone at UNLV exposed to a non-political history course. If there is still a one-course science requirement for all undergrads, I could see a course that is built around HOM, history of science, and history of technology as a good way to meet the requirement.

COMMENTATORS & ANALYSTS EXTRAORDINAIRE

Fourteen years ago, I started writing **Reflections,** a reincarnation of my life-long passion of writing all kinds of letters. The big difference was that that my reflective writings went public – from one to many, as I started sharing them with friends and relatives. And, from there, it went to their friends and relatives, and so on. Years ago, a student of mine created a blog, but seldom had I posted anything. I don't have a website either. It is all emails in a bcc - electronically old-fashioned mode.

I have several mailing lists and I am used to this inefficient mode of communication. I have Facebook and Twitter accounts too, but they too have remained unused. That is my approach to communication. Naturally, some of my readers write back and give comments. At times, a small dialog takes place. It has added clarity, expanded the topic, and sharpened my intellect – must at this stage of life.

For a number of reasons, not all the comments and commentators are included in the book– only those comments which are concise and strong. In reflective style of writings, inclusion of comments adds a new flavor. Initially, I never saved the comments. Also, sometimes, no comments were received. That is why the spaces following some **Reflections** are blank.

It is not merely a time to thank them, but also share a piece of 'immortality' that this book may bring! When I look at the credentials of these persons I am myself awed and wowed. These comments have come out of their incredible rich backgrounds. I don't think this list can be easily matched by any other author. Here are the names in some order:

Raju **Abraham**: Known for seven years. English professor – has taught in Baroda/India, Sana/Yemen, and presently in Oman with University of Nizwa, where I was a visiting professor for Spring-2009.

Francis Andrew is professor of English in the College of Applied Sciences, in Nizwa. His knowledge of Christian theology, astronomy and science in general is very extensive. He uses them effectively in writing science fiction books – nearly dozen. We met in 2009 and have been in regular touch since then.

Anand R. Bhatia, aged 71, is a retired professor of business from California State University, San Bernardino. We have known each other for 30+ years mainly for our common ideas and values, though we grew up in extremes cities – he, in Mumbai and me, in Bathinda. Anand is a very good story writer, but this flair is sacrificed for the time being for his love for real estate investments. I often tease him about marketing his voice and laughter, which are very deep and full.

Avnish Bhatnagar: My son, age 46, works at Google. His comments are fewer, but deep.

Rahul Bhatnagar: Distantly related - physician by training in India. He has an interesting job of medical director of drug safety with a pharmaceutical company; very astute commentator and analyst of nearly all my *Reflections*. He can refine an issue to a state that becomes undistinguishable from the one started with.

Larry Curnutt, (age 70) and I joined IU for PhD at the same time in the fall of 1968. I have told Larry story several times that he left IU after five years without PhD as his dissertation was not done. He joined a 2-year college community college and spent all his life there = being fully at peace with academic life. I often tell graduate student that for doctorate, right choice of supervisor is the most important gradient in the mix.

Irma Dutchie is our next-door neighbor, that I call her, our angel now after 12 years. Both physically and mentally, she can challenge any 60-year old. She has a Bostonian aura about her life style. Her love for life is as phenomenal as helping friends, strangers and charity organizations with her time, money and energy. She is my inspiration for life. At 84, she says that she can't be without a boyfriend!

Robert P. Gilbert (81) my PhD advisor and the last doctoral student before his left IU. He joined University of Delaware as Unidel Chair professor and retired at the age of 80. However, he remains active in supervising doctoral students researching in mathematical biology. We have remained in contact ever since. Many times, he tried to pull me into his old and new researches, but I was following my drummer within. His love for math has not affected his love for many other pursuits in life – that is inspiring.

Aaron Harris was the best student in the courses that he took from me and is about to get his PhD in math education while teaching fulltime in a high school and raising three kids.

Robert Meyer, 50+ years old was my student in several courses both as an undergraduate and graduate student. He suffered from a stroke in his 20s, later hit by a car in his wheel chair. He won the Governor's award for getting over the handicaps. He retired as a computer programmer from the Air Force. He loves to tutor math. He is an acclaimed poet of Nevada, and his poetry reminds me John Milton.

IBS Passi, 75 years old is working as an honorary professor of mathematics while continuing to do his research in algebra. He was a senior to me during our maser's from PU Chandigarh. He has the unique distinction of topping the exams both for his BA and MA. After PhD from UK, he taught for several years at KU and then moved to PU Chandigarh. We have been directly and indirectly in touch with each other ever since. He is a member of Indian Academy of Science, and has served as President of the IMS.

BhuDev Sharma 77: Known since 1990, math professor - taught in India, Trinidad and several universities in the US. He organized World Association of Vedic Studies and its biennial conferences in India and the USA. He is an able educational administrator. For several years, published the *vishwa vivek*, the first Hindi monthly magazine in the US.

Harpreet Singh: A rare combination of computer science, finance, active spirituality, and creative writing – always exploring and stretching his limits. He is 39 years old and known for 15 years – initially through his parents.

Sarvajit Singh, 75 and I have been known to each other since 1965 at KU. He is an established researcher in mathematical seismology and author. After years at KU, he moved to MDU Rohtak where he served as Professor and head of Math Dept and Dean before retiring there. He is settled in Delhi. He is a member Indian Academy of Science, and has served as President of the IMS.

Shankar and Sangeetha Venkatagiri is a couple in their 40s, who left their fulltime US jobs 13 years ago and went back to India. I got to know them through my son. Shankar got PhD in math from Georgia Tech, but now works for the IIM, Bangalore. Sangeetha is active in social work. They are fully adjusted in India with two kids – one adopted.

Subhash Sood: Physician by training in India, UK and USA - studied other systems of medicine too. He was deep into Scientology and established a center in Ambala Cantt, and translated several scientology books from English into Hindi. He suddenly died of a stroke in 2007 at the age of 73 - in a 100-year old dilapidated mansion in which he was born. He was my most avid reader and friend for over 25 years.

E. **Sooriamurthy**: Retired physics professor, Madurai University, India – known since 1968 – during our common PhD days at Indiana University, Bloomington. His son, Raja, computer science professor at Carnegie Mellon is an avid reader of my *Reflections* too. They are the second father-son duo and fan of my *Reflections*!